Fundamental Mathematics and Physics of Medical Imaging

Series in Medical Physics and Biomedical Engineering

Series Editors: John G. Webster, E. Russell Ritenour, Slavik Tabakov, and Kwan-Hoong Ng

Series in Medical Physics and Biomedical Engineering

Fundamental Mathematics and Physics of Medical Imaging

Jack L. Lancaster, Ph.D.
Research Imaging Institute
University of Texas Health Science Center at San Antonio, Texas, USA

Bruce Hasegawa, Ph.D.
Formerly of the University of California, San Francisco

CRC Press
Taylor & Francis Group
Boca Raton London New York

CRC Press is an imprint of the
Taylor & Francis Group, an **informa** business

CRC Press
Taylor & Francis Group
6000 Broken Sound Parkway NW, Suite 300
Boca Raton, FL 33487-2742

© 2017 by Taylor & Francis Group, LLC
CRC Press is an imprint of Taylor & Francis Group, an Informa business

No claim to original U.S. Government works

Printed on acid-free paper
Version Date: 20160701

International Standard Book Number-13: 978-1-4987-5161-2 (Hardback)

Library of Congress Cataloging-in-Publication Data

Names: Lancaster, Jack L., 1944- author. | Hasegawa, Bruce H., author.
Title: Fundamental mathematics and physics of medical imaging / Jack L.
Lancaster and Bruce Hasegawa.
Other titles: Series in medical physics and biomedical engineering.
Description: Boca Raton, FL : CRC Press, Taylor & Francis Group, [2016] |
Series: Series in medical physics and biomedical engineering | Includes
bibliographical references and index.
Identifiers: LCCN 2016029010| ISBN 9781498751612 (alk. paper) | ISBN
149875161X (alk. paper)
Subjects: LCSH: Diagnostic imaging--Mathematics. | Medical physics. | Imaging
systems in medicine.
Classification: LCC RC78.7.D53 L36 2016 | DDC 616.07/54--dc23
LC record available at https://lccn.loc.gov/2016029010

Visit the Taylor & Francis Web site at
http://www.taylorandfrancis.com

and the CRC Press Web site at
http://www.crcpress.com

Printed and bound in the United States of America by Publishers Graphics,
LLC on sustainably sourced paper.

Contents

Series Preface

THE SERIES IN MEDICAL PHYSICS AND BIOMEDICAL ENGINEERING describes the applications of physical sciences, engineering, and mathematics in medicine and clinical research.

The series seeks (but is not restricted to) publications in the following topics:

- Artificial organs
- Assistive technology
- Bioinformatics
- Bioinstrumentation
- Biomaterials
- Biomechanics
- Biomedical engineering
- Clinical engineering
- Imaging
- Implants
- Medical computing and mathematics
- Medical/surgical devices
- Patient monitoring
- Physiological measurement
- Prosthetics
- Radiation protection, health physics, and dosimetry
- Regulatory issues
- Rehabilitation engineering

- Sports medicine

- Systems physiology

- Telemedicine

- Tissue engineering

- Treatment

This is an international series that meets the need for up-to-date texts in this rapidly developing field. Books in the series range in level from introductory graduate textbooks and practical handbooks to more advanced expositions of current research.

The Series in Medical Physics and Biomedical Engineering is the official book series of the International Organization for Medical Physics (IOMP).

THE INTERNATIONAL ORGANIZATION FOR MEDICAL PHYSICS

IOMP represents more than 18,000 medical physicists worldwide and has a membership of 80 national and 6 regional organizations, together with a number of corporate members. Individual medical physicists of all national member organizations are also automatically members.

The mission of IOMP is to advance medical physics practice worldwide by disseminating scientific and technical information, fostering the educational and professional development of medical physics, and promoting highest-quality medical physics services for patients.

A World Congress on Medical Physics and Biomedical Engineering is held every three years in cooperation with the International Federation for Medical and Biological Engineering (IFMBE) and the International Union for Physics and Engineering Sciences in Medicine (IUPESM). A regionally based international conference, the International Congress of Medical Physics (ICMP), is held between world congresses. IOMP also sponsors international conferences, workshops, and courses.

The IOMP has several programs to assist medical physicists in developing countries. The joint IOMP Library Program supports 75 active libraries in 43 developing countries, and the Used Equipment Program coordinates equipment donations. The Travel Assistance Program provides a limited number of grants to enable physicists to attend the world congresses.

IOMP cosponsors the *Journal of Applied Clinical Medical Physics*. The IOMP publishes an electronic bulletin twice a year, *Medical Physics World*. IOMP also publishes *e-Zine*, an electronic newsletter, about six times a year. IOMP has an agreement with Taylor & Francis Group for the publication of the Medical Physics and Biomedical Engineering series of textbooks. IOMP members receive a discount.

IOMP collaborates with international organizations, such as the World Health Organization (WHO), the International Atomic Energy Agency (IAEA), and other international professional bodies, such as the International Radiation Protection Association (IRPA) and the International Commission on Radiological Protection (ICRP), to promote the development of medical physics and the safe use of radiation and medical devices.

Guidance on education, training, and professional development of medical physicists is issued by IOMP, which is collaborating with other professional organizations in the development of a professional certification system for medical physicists that can be implemented on a global basis.

The IOMP website (www.iomp.org) contains information on all its activities policy statements 1 and 2, and the "IOMP: Review and Way Forward," which outlines all activities and future plans of IOMP.

Preface

THIS TEXTBOOK IS A BASED ON THE BOOK *THE PHYSICS OF MEDICAL X-RAY IMAGING* by Bruce Hasegawa, Ph.D. There are two editions of this book: the first was published in July 1990 as a ring-bound soft copy and the second in November 1991 as a soft cover book, both by Medical Physics Publishing. Both have been out of print for over a decade. I used materials from Bruce's book for over 20 years in my advanced diagnostic imaging course in the Radiological Sciences Graduate Program at the University of Texas Health Science Center at San Antonio, TX. During this time, I substantially added to and improved the original content, which ultimately led to the new textbook *Fundamental Mathematics and Physics of Medical Imaging*. Mathematics was added to the title since many imaging and related concepts are formulated using calculus-based mathematics. "Medical X-Ray Imaging" in Bruce's book title was broadened to "Medical Imaging" since additional imaging modalities are now included (e.g., x-ray CT, MRI, and SPECT). The only common medical imaging modality not included is ultrasound, but many of the textbook's mathematical concepts apply. "Physics" was kept in the title because of its fundamental role in medical imaging, though many basic physics concepts are assumed to be covered in other courses (x-ray production, radioactive decay, interactions of radiation with matter, nuclear magnetic resonance, etc.). Basic physics concepts are extended in the textbook where needed to better describe the physical characteristics of imaging systems.

Acknowledgments

THIS TEXTBOOK WOULD NOT HAVE BEEN POSSIBLE WITHOUT THE EARLIER BOOK *THE PHYSICS OF MEDICAL X-RAY IMAGING* authored by Bruce Hasegawa, Ph.D. For many years, I communicated with Bruce concerning new and old material for the book. He eventually asked me to take over the book with the suggestion that I bring the new and old material together in a *new* book. I did this with the new textbook *Fundamental Mathematics and Physics of Medical Imaging*. Bruce passed in 2008, but I am confident that he would be pleased with what we have accomplished.

I have been developing ideas for this book for over two decades, and numerous people have helped. Some were students in my advanced diagnostic imaging course, who diligently worked out homework problems, helped to update figures, and discovered various problems with the book. Others were mentors who inspired me to ask "how" and "why" when answers were not obvious. I thank Francesca McGowen, physics editor, at CRC Press for her expert guidance through the process of developing and publishing this book. Finally, and most importantly, is my wife Maureen, who has supported and inspired me for more decades that I can mention.

Author

Jack L. Lancaster, Ph.D., is a professor of radiology at the University of Texas Health Science Center in San Antonio, where he has been teaching medical physics graduate level courses since 1989. He is the associate director of UTHSCSA's Research Imaging Institute and head of the Biomedical Image Analysis Division. His education is in both physics and medical physics, and he graduated in 1978 from the University of Texas Health Science Center at Dallas with a Ph.D. His experiences range from clinical nuclear medicine and radiology support at a major medical center in Dallas, TX, to training of technical staff and nuclear medicine residents at Brooke Army Medical Center, San Antonio, and finally to teaching and research at the UTHSCSA. Over the years, he has taught practically every course in the radiological sciences graduate program and mentored numerous students who now have successful careers as medical physicists.

Introduction

W HILE THIS TEXTBOOK IS APPROPRIATE FOR A GRADUATE LEVEL MEDICAL PHYSICS COURSE, we believe that it can also be used for an upper level undergraduate physics or engineering course. A reasonable prerequisite would be an introductory course covering basic medical imaging systems; however, experience has indicated that many students perform well without this prerequisite. Unlike Bruce's earlier book, this textbook includes a teacher's guide containing detailed answers to all homework problems. Also, example images from each imaging modality are provided for use with lectures and homework assignments. Display and analysis of example images are provided by the freely downloadable image processing application 'Mango' (http://ric.uthscsa.edu/mango/download.html or the publisher's website http://www.crcpress.com/product/isbn/9781498751612).

The textbook provides a structured approach for teaching advanced imaging concepts derived from basic concepts such as contrast, spatial and temporal resolution, and noise. As such, the book begins by introducing basic concepts (Chapters 1 through 3), moves to intermediate concepts (Chapters 4 through 8), and then to advanced concepts (Chapters 9 through 11). Following these sections, specific imaging methods (Chapters 12 through 13, dynamic x-ray imaging) and tomographic imaging modalities (Chapters 14 through 16, x-ray CT, MRI, and SPECT) are detailed. Homework problems keyed to each chapter are provided to monitor students' progress. The cover art for the textbook provides a wide assortment of terms that students will encounter. Though entering students may not be acquainted with many of these terms, they should be familiar with all of them after completing a course using the textbook.

I

Basic Concepts

CHAPTER 1

Overview

1.1 INTRODUCTION

This book is written assuming that you have a reasonable grasp of the physical mechanisms underlying the formation of medical images. Many of the basic physical concepts associated with medical imaging are fundamental to x-ray imaging, so numerous examples using x-ray imaging are provided. As such, students should have a basic understanding of the interaction of electrons with matter (bremsstrahlung) and the formation of x-rays, the interaction of x-rays with matter (photoelectric and Compton interactions), how the medical x-ray image is formed with radiographic film and intensifying screens, how (and in what units) the quantity of radiation is measured, and finally an appreciation of the principles of radiation protection. Additionally, since this book includes examples on nuclear medicine and magnetic resonance imaging, the student should have a basic understanding of associated physical concepts. There are several excellent texts that cover these topics, including those by Johns and Cunningham, Hendee, Ter Pogossian, Currey et al. (*Christensen's Physics of Diagnostic Radiology*), and Bushberg et al. (*The Essential Physics of Medical Imaging*). You are requested to consult these references if the aforementioned topics are not familiar to you.

1.2 IMAGING SYSTEM PERFORMANCE

The most basic features used to assess imaging system performance are *resolution* (spatial), *contrast*, and *noise*. Measurements of these basic features provide a foundation for studying and comparing medical imaging systems. Measures derived from pairs of these basic features provide further insight and include signal-to-noise ratio (SNR), modulation transfer function (MTF), and the Wiener spectrum. The most comprehensive measures of system performance incorporate all three basic features and include the Rose model equation (and related contrast detail analysis), and receiver operating characteristic (ROC) analysis.

Figure 1.1 illustrates the relationships between features and measures used to assess imaging system performance. Basic features (noise, spatial resolution, and contrast) are

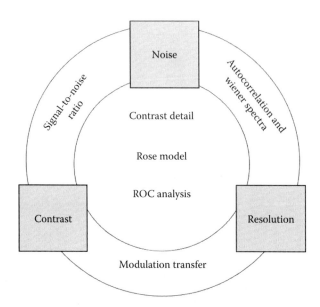

FIGURE 1.1 Schematic of medical imaging performance features ranging from basic (signal, noise, and contrast) to comprehensive (contrast-detail analysis, the Rose model, and receiver operating characteristic analysis).

indicated in boxes, and intermediate measures (SNR, Wiener spectrum, and MTF) are indicated in the regions bridging pairs of features. Finally, unifying measures (Rose model, contrast detail analysis, and ROC analysis) lie at the center of the diagram since they incorporate elements of all three basic features, implicitly if not explicitly. The diagram is not intended to suggest a hierarchy that one feature or measure is somehow more important than another. The student who contemplates Figure 1.1 should appreciate the unity of the science of medical imaging and understand these fundamental concepts as part of the whole, rather than as unrelated and independent pieces. To help understand this diagram, we provide both conceptual examples and basic quantitative methods.

1.3 BASIC PERFORMANCE FEATURES

1.3.1 Spatial Resolution

The spatial resolution of an imaging system can be intuitively defined in terms of the smallest spacing between two objects that can be clearly imaged. Spatial resolution varies widely between imaging modalities such that the spatial resolution of a conventional x-ray system with direct film exposure is approximately 0.01 mm while that of a CT scanner is approximately 1 mm. A satellite orbiting the earth's surface can record an object that is approximately 1 ft across. *In each case, the smallest distance between separate objects that a device can record is a measure of its spatial resolution.* This conceptual definition of spatial resolution is widely used in medical imaging. A more quantitative definition specifies spatial resolution of an imaging system in terms of its "point spread function" (PSF), which is the image of an "ideal" point object.

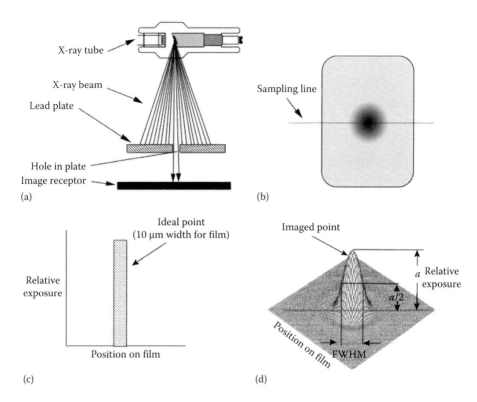

FIGURE 1.2 X-ray spatial resolution. (a) Schematic of x-ray image acquisition using a lead plate with a small hole, (b) developed x-ray image, (c) expected response for ideal imaging system, and (d) actual point spread function with the full width at half maximum width.

Figure 1.2 illustrates how the spatial resolution of an x-ray imaging system can be quantified using its PSF. In this example, the x-ray system images a "point" object, a small hole in an otherwise radio-opaque sheet of lead. X-rays pass through the small hole, forming an image of the point object. We can determine the 2-D PSF(x, y) of the system by recording the optical density (OD) values across the point image using a scanning optical microdensitometer if the image is recorded on a sheet of film, or by extracting digital values if the image is recorded digitally. These values are then converted to relative exposure values. A graph of the relative exposure is the PSF(x, y) (Figure 1.2d).

Ideally, one would see an exact representation of the object in the image, in that the width of the image would exactly match the width of the object. However, the image of a point object is always blurred by the imaging system. An index often used to indicate the extent of this blurring is the full width at half maximum (FWHM) of the PSF (Figure 1.2d). In Chapter 6, we describe how the PSF of an imaging system can be determined experimentally and describe additional approaches to characterize an imaging system's spatial resolution.

1.3.2 Image Contrast

Image contrast is broadly defined as the difference between adjacent regions in an image. In medical images, contrast refers to differences between neighboring tissues. For x-ray imaging, contrast between bone and soft tissue is high and contrast between fat and muscle is low. Radiographic contrast depends on several factors including the chemical composition of the object, the type of device used to record the image (whether it is film or an electronic detector), the energy spectrum of the x-ray beam, whether or not scatter radiation is present in the x-ray beam, and whether fog or some other baseline signal is present in the imaging device.

Image contrast is defined mathematically as "relative" contrast (Figure 1.3), here the fractional difference in x-ray exposure between the object and its surrounding background. In this figure, a small plastic disk is radiographed. Because of the nonlinear response of the film, we calculate image contrast using x-ray exposure (X). The background material (e.g., plastic platform) attenuates the x-rays passing through the background, while both the disk and the background material attenuate x-rays within the disk region. Thus, the x-ray fluence (photons/area) is higher outside of the plastic disk than it is beneath it, so the x-ray exposure is higher in the background than beneath the disk. If x-ray exposure

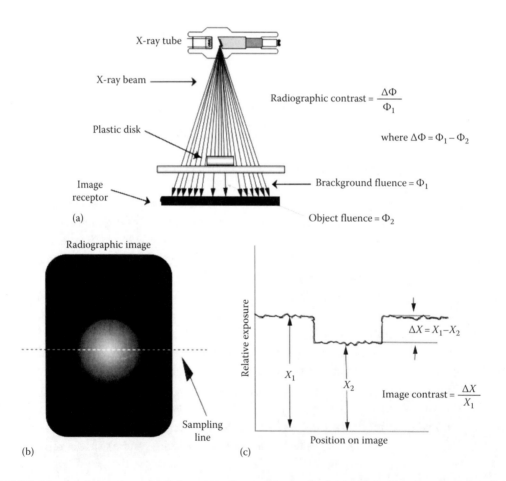

FIGURE 1.3   Image contrast. (a) Schematic of x-ray image acquisition for a low-contrast phantom (plastic disk), (b) developed x-ray image, and (c) graph of relative exposure levels for phantom and background.

in the "background" is X_1 and that for the disk is X_2, then the contrast of the disk relative to the background is

$$C = \frac{X_2 - X_1}{X_1} \qquad (1.1)$$

This contrast is unitless and can be positive or negative, though we often ignore the sign since it is understood from the context of the measurement. In the image given in Figure 1.3, x-ray exposure contrast is negative; however, film contrast is positive (Figure 1.3b). This contrast reversal is a property of x-ray films that produce negative images. We will discuss factors contributing to image contrast more thoroughly in Chapter 4.

1.3.3 Random Noise

Random noise relates to the uncertainty or imprecision with which a signal is recorded. In impressionist paintings, the artist would often create an image by painting a large number of small dots on the canvas. There is a great deal of uncertainty in the image being created when only a small number of dots have been placed on the canvas. As the number of dots increases, the precision or certainty with which the image is being represented increases. A similar thing happens in x-ray and nuclear medicine imaging. An image that is recorded with a small number of photons/area generally has a high degree of uncertainty or is noisy, while an image recorded with a large number of photons/area is more precise and less noisy. Grains in radiographic film, grains in intensifying screens, or electronic noise present in an electronic circuit or electronic detector can also contribute to x-ray imaging system noise. These factors contribute to the uncertainty or imprecision with which a signal is recorded, that is, to random noise.

An example of how random noise measured for x-ray imaging is given in Figure 1.4. Here, an image is acquired of a large plastic plate with a hole in it. The plate is suspended above the image receptor. A scanning microdensitometer is used to measure the density across the exposed radiograph and density converted to relative exposure. Ideally, we could repeat this experiment and obtain exactly the same densitometer readings each time. However, due to the random nature of emission of x-rays, attenuation within the plastic plate, and exposure of the film, this does not occur. Therefore, each density recorded has some level of uncertainty. This uncertainty is referred to as random noise. While there are several ways to quantify image noise, the most common noise measure is the *standard deviation* of image values in a uniform region of the image. In Figure 1.4, the standard deviation was assessed within the image region associated with the hole in the plate, but it could also be assessed in the uniform region outside of the hole. The concept of noise, how it is quantified, and its impact on images will be discussed in greater detail in Chapter 8.

Imaging science would be simple if an imaging technique could be described, and its performance quantified, in terms of only one of the three basic features: noise, spatial

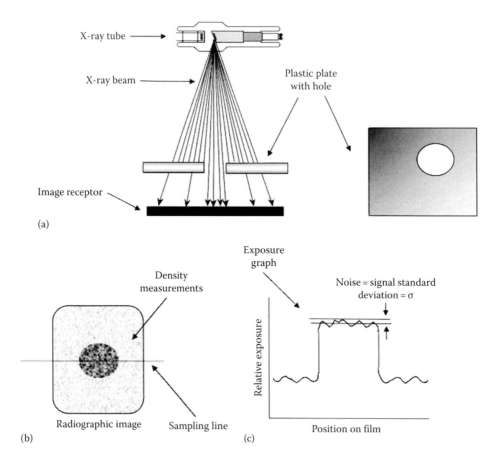

FIGURE 1.4   Concept of noise. (a) Schematic of x-ray image acquisition using plastic plate with hole, (b) cartoon of x-ray image, and (c) graph of relative exposure illustrating method to measure noise as a standard deviation of exposure.

resolution, or contrast. In such a simple world, one could say that a screen/film combination has better spatial resolution than a CT or MR image and therefore is better for all medical imaging applications, that a CT scanner is better than a film/screen system because it provides higher bone-to-soft tissue contrast, or that MRI is better than CT because of its high within soft tissue contrast. These simple statements obviously are not true for all cases and the imaging modality of choice depends on specific medical imaging needs (i.e., broken bone vs. brain tumor). To continue this line of thought, a nuclear medicine bone scan can reveal metastases to bone before radiographic or MRI changes are detectable. To evaluate the suitability of a specific system for a given imaging task, scientists must understand the tradeoffs in the design of the instrumentation used to image the human body as well as the needs of the imaging task. Several intermediate measures based on pairs of the basic features (noise, spatial resolution, and contrast) help in evaluating imaging system performance.

1.4 INTERMEDIATE PERFORMANCE MEASURES

1.4.1 Modulation Transfer Function

1.4.1.1 Spatial Resolution and Contrast

It is not sufficient to specify the capability of an imaging system only in terms of the smallest size of objects that can be distinguished. Detectability is also affected by the relative contrast of objects. In fact, for all medical imaging systems, the contrast with which an object is imaged decreases as the object size diminishes due to blurring. The relative contrast between objects literally disappears when their spacing diminishes toward the FWHM of the system PSF. A measure that incorporates both contrast and resolution is the MTF. The MTF provides contrast information as a function of diminishing object size, not just a limiting size. The MTF(f) is sometimes assessed using periodic objects that appear sinusoidal to the imaging system, as in Figure 1.5. Higher frequencies relate to smaller distances between the light and dark regions in this pattern, and the diminished contrast at higher frequencies (Figure 1.5c) leads to diminished spatial resolution. The scientific and mathematical basis of the MTF(f) will be covered in Chapter 6, with examples from several medical imaging systems.

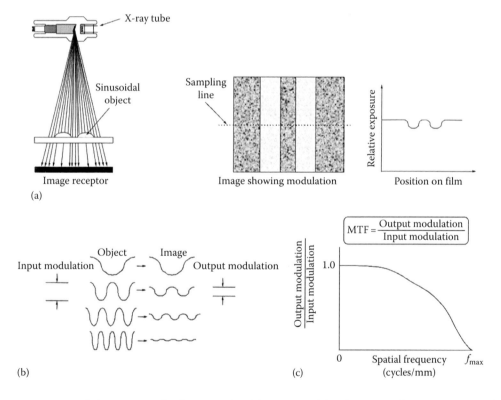

FIGURE 1.5 Modulation transfer function (MTF) concept. (a) Schematic of x-ray exposure, the sinusoidal phantom, and resulting relative exposure. (b) Graphic illustrating how the MTF could be measured using phantoms representing different spatial frequencies. (c) Example of a continuous MTF curve and equation for modulation.

1.4.2 Signal-to-Noise Ratio

1.4.2.1 Noise and Contrast

Another important measure concerning imaging system performance is SNR, a familiar term to electronic engineers or scientists working with low-level signals in electronic circuits. As the terminology implies, SNR is the ratio obtained when the image's signal is divided by the image's noise, both measured using the same units (Figure 1.6). The SNR is therefore a unitless index. For imaging systems, the signal of interest is the difference between the object of interest and its surrounding background (ΔX in Figure 1.6), and the noise is the uncertainty with which the object is recorded (the standard deviation = σ). For example, if one were imaging a tumor that was found in the liver, the signal would be the difference between the tumor and the surrounding tissue. In this example, the noise could be assessed as the standard deviation in the nearby surrounding tissue. The ratio of these two numbers would be SNR. The SNR in medical images will be further investigated in Chapter 8.

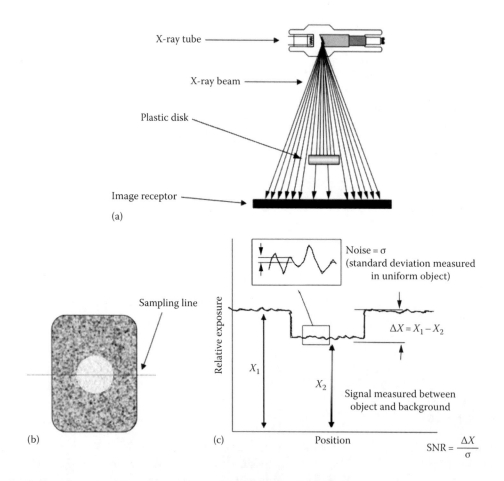

FIGURE 1.6 Signal-to-noise ratio (SNR). (a) Schematic of a setup for x-ray exposure to assess the SNR. (b) Graphic of resulting exposure with line for assessing responses. (c) Graph of relative exposure vs. position used to assess the signal, the noise, and the SNR.

1.4.3 Wiener Spectrum

1.4.3.1 Noise and Spatial Resolution

Another intermediate measure that combines noise with spatial resolution is the Wiener spectrum (Figure 1.7). Just as the response of an imaging system decreases as the object's spatial dimension becomes smaller and smaller, the response of the system to noise fluctuations decreases as the spatial extent of noise becomes smaller and smaller. The theoretical frequency spectrum of random noise is uniform across a wide range of frequencies, so the system frequency response to random noise will be similar to the system MTF(f). The Wiener spectrum describes noise power as a function of spatial frequency and equals the Fourier transform of the autocorrelation function in a uniformly exposed radiographic image. This topic will be explored further in Chapter 9.

Finally, there are image quality measures that attempt to incorporate all three basic image features (*spatial resolution, contrast*, and *noise*) as well as the performance of the observer in the evaluation of an imaging system. These methods include the Rose model, related contrast-detail analysis, and receiver operating characteristic (ROC) analysis.

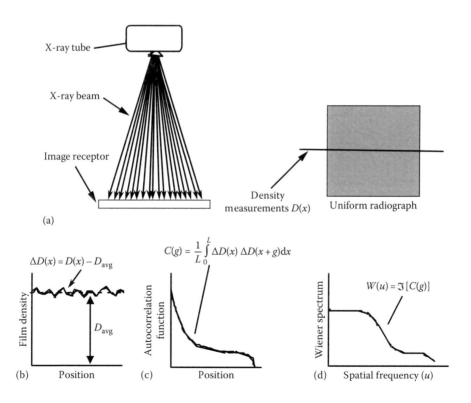

FIGURE 1.7 Concept of Wiener spectrum. (a) Schematic of x-ray image acquisition for used in determining the Wiener spectrum. (b) Shows how delta signal value data are to be calculated. (c) The intermediate calculation of the autocorrelation function. (d) The calculation of the Wiener spectrum as the Fourier transform of the autocorrelation function.

1.5 ADVANCED PERFORMANCE MEASURES

1.5.1 Rose Model

The Rose model is a mathematical relationship (Equation 1.2) between the SNR (k) at the threshold of detectability, object size (A), and contrast (C):

$$k^2 = C^2N = C^2\Phi A \tag{1.2}$$

where
  k is the SNR needed to just see an object in an image
  C is the contrast of the object with respect to surrounding background
  N is the number of photons used to image the object of area A (usually in background)
  Φ is the photon fluence (N/A) used to form the image
  A is the area of the object

The Rose model relates the three fundamental features using "size or A" for spatial resolution, Φ for noise, and C for contrast. The equation can be used to predict whether an object with a given set of characteristics (size and contrast) can be visually detected in an image created at a certain noise level. A value of k (SNR) in the range of 5–7 has been reported to be adequate for many imaging tasks. Using Equation 1.2, with an assigned value for k, we can estimate the size of the smallest object (A) that we might be able to see at contrast level (C) and photon fluence (Φ). The Rose model is a key element in our assessment of the ability to perceive low-contrast objects in a noisy image, conditions often found in radiology. We will further describe and expand on the Rose model in Chapter 9.

1.5.2 Contrast Detail Analysis

A Rose model phantom can be used for contrast detail analysis (Figure 1.8a). An observer views an image of the phantom with object size changing in one direction and contrast changing in the other. The observer reports the size of the smallest object they perceive at each contrast level. The result of such an analysis is a *contrast detail* curve in which the size (i.e., detail) of smallest observable objects is plotted against their contrast. A different curve can be plotted for each noise level (Figure 1.8c). These curves illustrate the constraints on signal detection implied by the Rose model equation. For example, smaller objects must have higher contrast to be seen in the image, approaching the *resolution-limited* region of the graphs. Imaging with low photon fluence (low SNR) requires higher contrast than using high photon fluence (high SNR). Finally, the disparity between low and high SNR tends to increase with increasing object size toward the *noise-limited* region of the graphs. Such a set of curves, one for each noise level, can be used to provide a relationship between the contrast and object size needed under differing noise conditions.

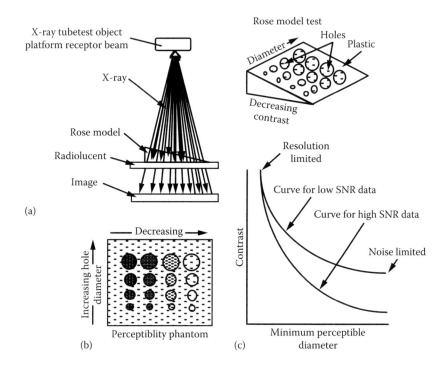

FIGURE 1.8 Contrast detail curves. (a) Schematics of the x-ray image acquisition and the Rose model phantom. (b) Graphic of the x-ray image acquired using the Rose model phantom. (c) Contrast detail curves derived using the Rose model phantom for high and low SNR imaging.

1.5.3 Receiver Operating Characteristic Analysis

Receiver operating characteristic curve (ROC) analysis is considered the definitive test of a diagnostic imaging system (Figure 1.9) since it can include observers. It can be used to evaluate imaging systems, imaging techniques, and even image analysis methods. In ROC analysis, the *true-positive fraction* (e.g., the fraction of patients correctly stated as having the disorder) of a diagnostic test is plotted against the *false-positive fraction* (fraction of patients incorrectly stated as having the disorder). The true-positive fraction is also called the "sensitivity" of a test and relates to how well a test performs in detecting a disorder.

Unlike other performance measures, ROC analysis does not explicitly use noise, contrast, or resolution as dependent or independent variables, but outcomes are dependent on these factors. This makes it easier to compare different imaging systems to determine which might be best for a particular diagnostic imaging task. An ideal system would give no false positives unless the observer insisted upon calling everything positive, and its ROC curve therefore would pass through the upper left corner of the graph. On the other hand, if the image conveyed no information and the observer was forced to guess whether or not the object was present, the ROC curve would be a diagonal line from the lower left to the upper right corner. Therefore, the amount by which the ROC curve bows away from the diagonal and toward the upper left-hand corner is a measure of the usefulness of the imaging technique. More details concerning ROC analysis methods will be presented in Chapter 11.

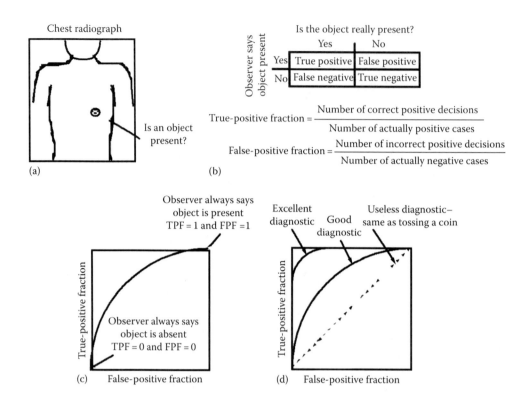

FIGURE 1.9 Receiver operating characteristics (ROC) analysis. (a) Graphic of radiograph of patient with or without lesion to be evaluated by multiple observers. (b) Four possible outcomes (true positive, false positive, true negative, and false negative). (c) Example of graphing the ROC curve as a true-positive fraction of responses (TPF) vs. false-positive fraction of responses (FPF). (d) Graph of ROCs indicating excellent, good, and useless diagnostic capabilities.

1.6 ORGANIZATION OF THE BOOK

The book is organized into five sections with 16 chapters: Sections I through III (Chapters 1 through 11) cover basic, intermediate, and advanced math and physics concepts, while Sections IV through V (Chapters 12 through 16) cover several specific medical imaging modalities (X-ray CT, SPECT, and MRI).

HOMEWORK PROBLEMS

P1.1 Assume that we have taken a radiograph of a small circular plastic object from which we have made the following relative exposure measurements along a line including the object: 359, 376, 421, 424, 394, 371, 423, 349, 399, 346, 482, 476, 498, 501, 528, 449, 501, 530, 525, 439, 502, 467, 521, 520, 523, 479, 528, 529, 476, 523, 430, 392, 439, 390, 429, 439, 387, 380, 420, 429

   (a) Make a graph of these values as a function of position.

   (b) Calculate *contrast, noise,* and the *signal-to-noise ratio.*

   In each case, define any ambiguous terms and describe how you determined each value.

P1.2 Assume that a PSF is described by the following Gaussian function:

$$f(x) = \frac{1}{\sqrt{2\pi}\sigma} e^{\frac{-(x-\mu)^{-2}}{2\sigma^2}}$$

(a) If $f(x)$ represents a probability density function on the domain $(-\infty$ to $+\infty)$, show that μ is the mean value and σ is the standard deviation.

(b) Show algebraically that the FWHM for a Gaussian PSF is

$$\text{FWHM} = 2\sqrt{2\ln(2)}\sigma$$

P1.3 The following are image signal measurements taken at equal intervals across the image of a point object:

Position	0	1	2	3	4	5	6	7	8	9	10	11	12	13	14	15	16	17	18	19	20
Value	0	1	3	7	14	25	41	61	80	95	100	95	80	61	41	25	14	7	3	1	0

(a) Calculate the standard deviation of the profile, and then use the result from (2b) to estimate the FWHM of the PSF.

(b) Graph the signal measurements given in the table and use this graph to estimate the FWHM. Compare the result with that in (a).

P1.4 Given film density values for a PSF, explain how you would convert these values to relative x-ray exposure.

P1.5 Conceptually, what does the MTF represent?

CHAPTER 2

Medical Imaging Technology and Terminology

T HE READER SHOULD HAVE an understanding of the basic principles of conventional radiography in which photographic methods capture radiographic images. Typically, for planar x-ray imaging, a patient is positioned between an x-ray source and a film/screen cassette. The cassette contains an intensifying screen that, when exposed to radiation, emits light that exposes the film. The film is developed and viewed by a radiologist. A film is a multipurpose component used for image acquisition, viewing, and archiving. Increasingly planar radiographs are captured, processed, displayed, and stored digitally rather than using film.

Medical images are acquired using both digital planar and tomographic imaging systems, and this chapter briefly summarizes methods used for both (Figure 2.1). Additional details regarding these imaging systems will be covered in later chapters, with focus on imaging performance.

2.1 DETECTORS

2.1.1 Image Intensifier

Planar radiographs provide high-resolution images of stationary objects that are used in many clinical studies, but "dynamic" images of moving body parts are needed for many studies. This is seen in the circulatory system where a series of x-ray images must be acquired to monitor cardiac motion, image or measure the flow of a contrast agent through the circulatory system, or map temporal changes associated with other organs.

In the "olden days," radiologists used fluorescent screens to observe the motion of body parts under x-ray examination (Figure 2.2). An x-ray fluorescent screen is one that emits light when exposed to x-rays. The patient's body was positioned between the screen and the x-ray tube. X-rays passed through the patient's body and struck the screen, allowing the radiologist to view the image from the fluorescent screen.

2.1. Detectors
- 2.1.1. Image Intensifiers
- 2.1.2. Scintillators
- 2.1.3. Gas Filled
- 2.1.4. Solid State

2.2. Digital Planar Imaging
- 2.2.1. Film Digitization
- 2.2.2. Photostimulable Phosphors (PSP) Systems
- 2.2.3. Scanning Detector Arrays
- 2.2.4. Digital Radiographic (DR) Systems

2.3. Computed Tomographic Imaging
- 2.3.1. Positron Emission Tomography (PET)
- 2.3.2. Single-Photon Emission Computer Tomography (SPECT)
- 2.3.3. X-Ray Computed Tomography (CT)
- 2.3.4. Magnetic Resonance Imaging (MRI)

FIGURE 2.1 Detectors, digital planar imaging methods, and computed tomographic imaging methods used in diagnostic medical imaging.

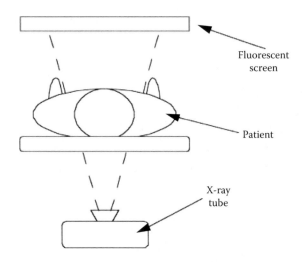

FIGURE 2.2 Key components of a conventional x-ray fluorography system.

There were three basic problems with this technique. First, the resulting image was very dim. The radiologist had to view the image in a darkened room. In fact, the radiologist had to stay in the darkened room, or wear dark red glasses, so that their eyes remained dark-adapted. Second, the radiologist usually looked directly at the fluorescent screen, so his or her head was in the path of the x-ray beam. This problem could be overcome by viewing the image through one or more mirrors, but this was an annoying complication. Finally, the image was so dim that it could not be properly recorded using a movie or video camera.

In the late 1940s and early 1950s, the x-ray image intensifier was introduced as a way to eliminate both the dim fluorescent image and the requirement for direct viewing. The image intensifier provided a means to both image and record dynamic processes in the body.

In practice, the radiologist moved the image intensifier to position it over the body part of interest. For example, an abdominal radiologist might use an image intensifier system to examine the gastrointestinal tract of a patient suffering from an ulcer. The patient was given a barium contrast agent either taken orally (or introduced into the GI tract by some other route). The radiologist viewed the lining of the GI tract after it was coated with the contrast agent to find areas where an ulcer might be located.

As its name suggests, the purpose of the image intensifier was to amplify the light image produced by the fluorescent screen to generate a much brighter image for viewing. The image formation process begins when x-rays strike the input screen of the image intensifier or II (Figure 2.3). The II's fluorescent screen converts the x-ray image into a dim light image, similar to a conventional fluoroscopy screen. Signal amplification begins at the photocathode that converts the light image into a 2-D pattern of electrons. The electrons are accelerated toward the anode end by a high voltage in a way that preserves the geometry of the image. The electrons are focused onto a smaller output phosphor screen where their energy is converted back into light. The combination of electron acceleration and geometric mini-fication produces a light image at the output phosphor that is 1000–5000 times brighter (photons/area) than the light image produced by the input fluorescent screen. As a result, the image intensifier increases the brightness of the x-ray image (called the fluoro image) to a level that can be readily monitored by a video camera.

The fluoro image is viewed on a video monitor by the radiologist, and a video recorder similar to those used in homes to record football games and our favorite television

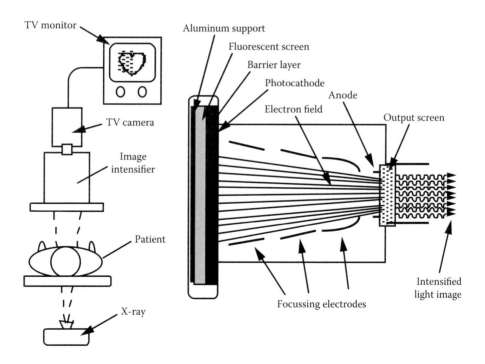

FIGURE 2.3 Added components of an image intensifier–based x-ray fluorography system with image intensifier detailed.

programs can be used to record the study for later review. Alternatively, the fluoro image can be recorded with a movie or cine camera on film with better spatial resolution and better contrast than video, but which must be chemically processed.

Most importantly, the image intensifier produces a "live" fluoro image where the radiologist can, for example, watch dynamic changes such as ventricular contraction in the heart. The image intensifier can also be moved across the patient to survey a large region of anatomy. This is useful in abdominal radiography where radiologists must locate a specific region of the patient's internal anatomy prior to recording high-resolution static planar images (called spot films).

2.1.2 Scintillator

The scintillation detector (Figure 2.4) has been used in positron emission tomography (PET) systems, nuclear medicine gamma cameras, single-photon emission computed tomography (SPECT) systems, and older x-ray computed tomographic (CT) imaging systems. A scintillator is a material such as NaI(Tl) or cadmium tungstate that emits light when exposed to radiation, where the brightness of the emitted light is proportional to the energy absorbed. The light signal from the scintillator is converted into an electrical signal by mounting the scintillator on a photomultiplier tube or a sensitive photodiode. Scintillation crystals have a high average atomic number and high density, making them excellent absorbers of x-rays. The thickness of the crystal can be increased to ensure nearly 100% absorption of x-rays and gammas. These detectors generate an analog electronic signal proportional to the brightness of the light generated by the scintillator. The resulting signal is delivered through an electronic amplifier to an analog-to-digital converter (ADC) that converts the voltage signal to a digital value that is processed and stored for later use.

Unlike an image intensifier that produces an image over a large area, a scintillation detector captures the radiation intensity only at the location of the scintillation crystal. The gamma camera used in nuclear medicine departments is an exception to this rule. It uses a large single crystal with an array of photomultiplier tubes attached to determine where in the crystal gamma radiations are absorbed. Scanning a small scintillation detector with a focused collimator to record the intensity at each point has also been used to generate an image, similar to older nuclear medicine rectilinear scanners. Medical x-ray CT imaging

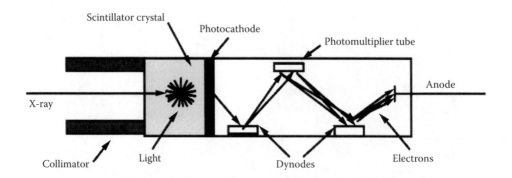

FIGURE 2.4 Scintillation detector with collimator, NaI(Tl) crystal, and photomultiplier tube.

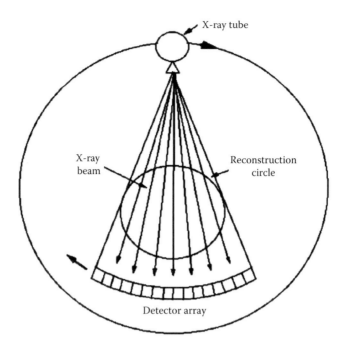

FIGURE 2.5 Third-generation x-ray CT system showing rotational movement of x-ray tube, the array of detectors, and the limits of the reconstructed image (scan circle).

systems use an array of detectors arranged in a circular arc (Figure 2.5). This detector array is rotated about and translated along the patient to collect sufficient data to reconstruct tomographic images of the patient volume within the scanned range.

2.1.3 Gas Filled

Another detector used in digital radiographic (DR) systems is the gas ionization detector (Figure 2.6). It was used in older CT scanners as an alternative to scintillation detectors. A gas ionization detector can be thought of as a volume of special gas contained in an enclosure having one face that is transparent to x-rays that serves as the radiation window. The detector contains two electrodes, one at a negative electrical potential and one at a positive electrical potential. An x-ray enters the detector through its radiation window and dislodges electrons from the gas molecules forming pairs of ions. The resulting charges are collected by the electrodes and form an electrical current in the detector electronics that is proportional to the total number of ion pairs produced. Like scintillation detectors, this device measures radiation exposure at the location of the detectors. Therefore, these detectors were arranged in a circular arc and rotated about the patient for CT scanning (Figure 2.5).

The advantage of the gas ionization detector is that it is inexpensive and can be made relatively compact. The disadvantage is that gases are poor x-ray absorbers compared to solid crystalline scintillators. Xenon is used in gas ionization detectors since it has excellent x-ray absorption in comparison to other gases due to its relatively high atomic number (high-z). Additionally, to further increase x-ray absorption, the xenon gas is pressurized to several atmospheres (increasing density).

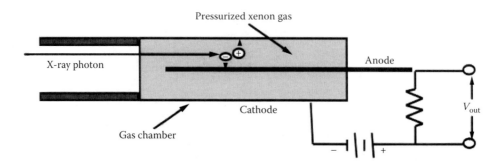

FIGURE 2.6 Gas-filled ionization detector with collimator, gas chamber, and associated electrical components.

2.1.4 Solid State

The most extensive use of solid-state detectors in medical imaging is with DR imaging panels. These panels consist of a large array of closely packed detectors, with each detector assigned a row–column address and corresponding physical location. The detectors are classified as either "indirect" or "direct" based on how they convert absorbed x-ray energy into an electrical signal. Indirect-conversion detectors are hybrids as they contain both scintillators and semiconductors, and conversion is a two-step process. In the first step, absorbed x-ray energy is converted to light by the scintillator(s), and in the second step, the light is converted to an electrical signal by a photodiode and stored on a capacitor. Direct-conversion detectors both absorb x-ray energy and convert it to an electrical signal that is stored in a capacitor. Row–column readout electronics for the digital panels encodes locations of each detector and converts stored electrical signals into digital signals.

The efficiency of x-ray absorption is generally higher for indirect-conversion detectors since its CsI:Tl scintillators have a higher average atomic number than the detection component in direct-conversion detectors (amorphous selenium [a-Se]). The dynamic range of solid-state detectors is high ($\sim 10^4$), necessitating a 16-bit integer to cover the range. Unlike film-screen systems, solid-state detectors have a linear response across their dynamic range so that they are useful across a wider range of x-ray exposures, supporting lower or higher exposures without overexposing or underexposing as seen in film/screen systems.

2.2 DIGITAL PLANAR IMAGING

Film–screen imaging has been the mainstay for planar x-ray imaging for many decades, and similarly image intensifiers have been the mainstay for dynamic imaging. Various digital imaging technologies including (1) photostimulable phosphor systems, (2) scanned detector arrays, and (3) digital flat panels are replacing these older technologies. A picture archiving and communicating system (PACS) can provide rapid access to and viewing of the digital images. X-ray films can be entered into a PACS using film digitizers to convert them to a digital format.

2.2.1 Film Digitizer

X-ray film digitization is a technique in which the radiographic film is placed in an instrument that scans the film to generate an electronic signal proportional to its optical density as a function of position. A typical system is shown in Figure 2.7. In this instrument, a narrow-beam light source (usually a laser) is projected through the film, and the intensity recorded with a photomultiplier tube. The optical density (OD) of the film is defined as

$$OD = -\log(\text{transmittance}) = \log\left(\frac{I_o}{I}\right) \tag{2.1}$$

where
I_o is the incident light intensity (w/o film)
I is the transmitted light intensity

This relative measure corrects for differences in light source intensities between scanners and within-scanner changes over time. The laser film scanner produces an electronic signal that is proportional to film density that is digitized and stored using 10–12 bits of a 16-bit integer, with images that are formatted into a 1024 × 1024 or finer pixel matrix.

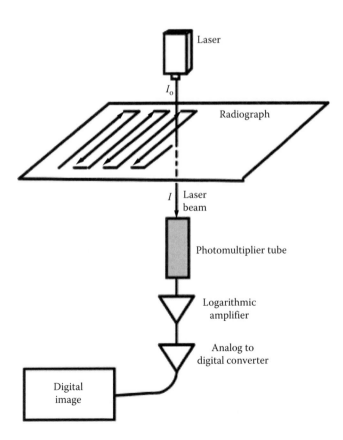

FIGURE 2.7 Laser film scanner used for film digitization.

2.2.2 Computed Radiographic (CR) Systems

A digital imaging alternative to the film-screen system, based on a photostimulable phosphor screen, was introduced in the 1980s (Figure 2.8). A photostimulable phosphor screen emits light when stimulated by a light of a different wavelength (hence photostimulable). When exposed to x-rays, about half of the energy in the x-ray beam is trapped in the photostimulable screen as a "latent image" that decays slowly over time. This energy is released by scanning a red laser beam across the plate, generating blue light (the luminescence). The brightness of the luminescence is proportional to the intensity of the x-rays originally striking the screen. The red light is filtered out, and the blue light is amplified and converted by a photomultiplier system to an electrical signal representing x-ray intensity. The electrical signals are digitized and using digitally encoded scanner positions converted to a 2D digital image.

Photostimulable phosphor screens have several advantages over film/screen systems. First, similar to solid-state detectors, photostimulable phosphors are less prone to underexposure or overexposure, because of their wide dynamic range of radiation exposure (10^4–10^5) compared to film (~10^2). This is important for portable or emergency chest radiography where the detector must record high exposures behind the lung simultaneously with low exposures through the abdomen. Second, the response of photostimulable phosphors is linear, a property that simplifies relating pixel values to x-ray exposure (Homework Problem P2.4). Finally, the image is obtained electronically, without chemical processing, and the photostimulable screens are reusable. Importantly, photostimulable phosphor screens are designed for mounting in conventional size cassettes and served as

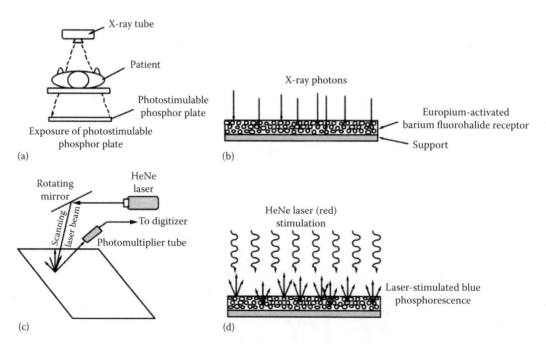

FIGURE 2.8 Photostimulable phosphor system general layout (a), formation and stimulation of the latent image (b), and scanning used during stimulation by laser and readout (c, d).

a DR replacement for film/screen cassettes prior to updating to DR plate–based systems. Among the disadvantages of photostimulable phosphor systems are their cost ($500,000 or more) and the slight degradation in spatial resolution in comparison to film-screen systems, which may be critical in diseases such as the detection of pneumothorax or diffuse interstitial disease.

Similar to indirect DR detectors, the photostimulable phosphor systems use a two-step process to formulate a digital image. However, since the digital image in these systems is not read out at the time of exposure, it is not categorized as a DR system but rather as a CR system. Regardless of the categorization, both DR and CR systems only produce digital images.

2.2.3 Scanning Detector Array

Arrays of scintillation, gas ionization, or solid-state detectors can be used in scanned projection systems. In these systems, the radiation beam is collimated using fan-beam or pencil-beam geometry (Figure 2.9). The x-ray beam passes through the patient and onto the detectors that are arranged in a linear array or circular arc. Scanning the x-ray beam and the detector array across the patient forms an image. The detectors accommodate a wide range of x-ray exposures making them less prone to overexposure or underexposure than film/screen systems. In comparison with broad beam geometries, the narrow x-ray beam greatly reduces scatter, thereby improving the detection of subtle differences in x-ray absorption by soft tissues in the patient. Scanned detector array systems have not been

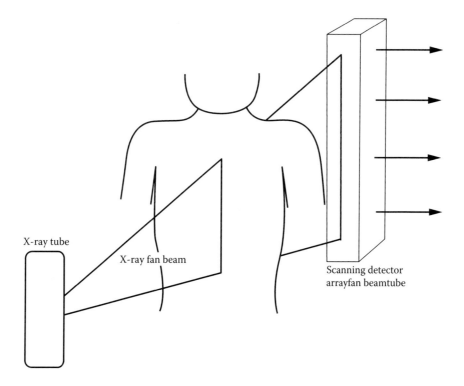

FIGURE 2.9   Fan beam x-ray scanning system with vertically aligned array of detectors.

widely adopted for routine clinical studies because they have poorer spatial resolution than film/screen and photostimulable phosphor systems, are susceptible to motion artifacts, and are expensive to purchase and maintain in comparison to conventional film/screen systems. Finally, x-ray CT can be used as a scanned detector array system, as exemplified by the scout scan used to select scan range along the body.

2.2.4 Digital Radiographic Systems

DR systems have emerged as a key component of picture archiving and communications systems (PACSs) in radiology departments. DR systems are based on digital flat panels containing a large array of small solid-state detectors. Unlike PSP systems, DR systems do not require physical scanning since pixel locations are directly linked to detector locations within their flat panels. Exposure times are similar to film/screen and photostimulable phosphor systems, but digital images are available immediately following the x-ray exposure. Various interfaces to a PACS are available for DR imaging systems including optical fiber, Ethernet, and Wi-Fi.

Digital flat panels were developed in the mid-1990s as replacements for film/screen systems. Initially, the panels were restricted to static imaging due to readout times. However, their use in dynamic imaging, where images must be acquired rapidly, soon followed as panel readout time was reduced to acceptable limits. Further advancements in DR system design have led to replacement of image-intensifier systems in special procedures and cardiac catheterization labs, supporting both dynamic and static planar imaging needs. Advantages of DR systems over image intensifiers for dynamic imaging include (1) absence of geometric distortion, (2) uniform response across the image, and (3) direct conversion to a digital image.

2.3 MULTIPLANAR IMAGING: COMPUTED TOMOGRAPHY

The older terminology for computed axial tomography (CAT) was shortened to CT. Any approach that computes tomographic section images (*tomo* means cut) can be classified as CT. CT therefore includes several modern radiological imaging methods: x-ray CT, SPECT, PET, magnetic resonance imaging (MRI), and even some forms of ultrasound imaging. As implied by their computing nature, all CT images are digital.

The basic math describing formulation of a 2D section image from projections taken about an object has been known since the early 1900s. However, technological limitations (mostly computer) hampered the development of CT systems until the 1960s. There are two major categories of CT systems: (1) emission computed tomographic (ECT) systems that form images from radiations emitted from the body, and (2) transmission computed tomographic (TCT) systems that form images from radiations transmitted through the body.

2.3.1 Positron Emission Tomography

The earliest ECT imagers are PET systems based on the annihilation radiation originating from the site of positron interactions. The detection of the pair of annihilation photons determines the line along which the photons passed through the body and serves as the basis for

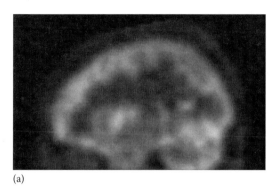
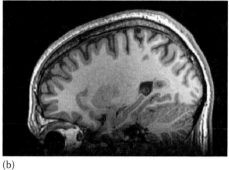

(a)   (b)

FIGURE 2.10   PET F-18 FDG image of brain (a) and corresponding anatomical MRI of the head (b). The FDG is a functional image with pixel values representing glucose consumption, while the high-resolution MRI provides anatomical detail.

forming projections. Research PET imagers were developed in the 1960s and commercial units soon followed. Positron-emitting radioisotopes of elements commonly found in biological molecules, particularly O-15 and C-11, were key to the acceptance of PET. Of major importance was the use of O-15 water for cerebral blood flow, the basis of functional brain research since the late 1970s. Finally, the formulation of F-18 as fluorodeoxy-D-glucose (FDG) provided a means to assess glucose metabolism in the brain and glucose metabolic activity of cancers within the body (Figure 2.10). The FDA approved the use of FDG for cancer imaging, and many cancer centers use PET/CT systems for treatment planning and monitoring of results.

2.3.2 Single-Photon Emission Tomography

The term SPECT was chosen to clearly distinguish this ECT system from PET since it does not use two photons. The most popular SPECT approach, using a rotating gamma camera, was introduced in the 1970s followed by many improvements in the embedded software and hardware. A problem with SPECT is attenuation correction, which depends on the distribution of the radiotracer within the body, and this is not well known until the SPECT image is determined. Modern SPECT systems therefore deal with this problem using iterative reconstruction methods. An advantage of SPECT over PET systems is the use of long-established radiotracers based on Tc-99(m) and cost. An example of this is the use of Tc-99(m) ethyl cysteinate dimer (ECD) to localize the difference in activity during (ictal) and between (interictal) seizures (Figure 2.11).

2.3.3 X-Ray Computed Tomography

X-ray CT was introduced in the 1970s and rapidly became the method of choice for imaging of the brain and spinal cord where the surrounding bony structures confounded viewing of the brain's soft tissues using planar methods. X-ray CT is classified as a TCT system as it is based on the x-rays that pass through the body. Early x-ray CT systems were limited by spatial resolution and speed, but each decade has seen rapid improvements, and sub-mm resolutions and multislice detector systems with high scan speeds have emerged.

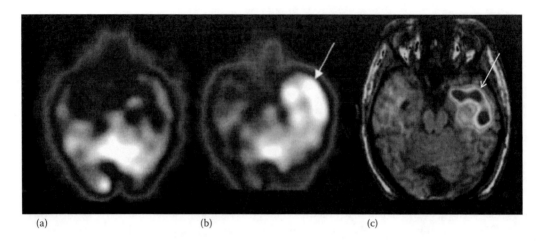

(a)   (b)   (c)

FIGURE 2.11   SPECT Tc-99(m) ECD imaging of interictal (a), ictal (b), and ictal minus interictal (c) overlaid onto MRI to highlight focal seizure activity. Adapted from Lewis, P.J. et al., *J. Nucl. Med.*, 41, 1619, 2000, Figure 4. With permission.

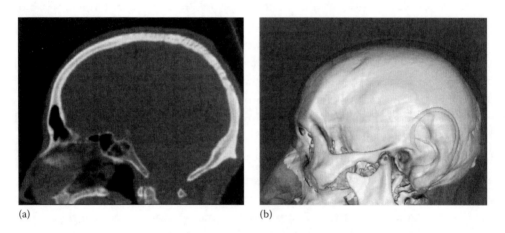

(a)   (b)

FIGURE 2.12   Midsagittal x-ray CT section image (a) provides high contrast between bone and soft tissue but little contrast within the soft tissues within the brain and surrounding the skull. The surface image of bone and skin (b) made from CT sections is geometrically accurate and can be used for measurements about the head or bones.

The high-quality bone imaging and exceptional bone/soft tissue contrast with x-ray CT provide a means to render highly accurate geometrical models of the body (Figure 2.12). The advent of MRI in the 1980s, with its exceptional soft tissue contrast, reduced the need for x-ray CT. However, x-ray CT continues to be an exceptional imaging system due to its geometrically accurate images and its standardization of tissue signals based on CT numbers.

2.3.4 Magnetic Resonance Imaging

The nuclear magnetic resonance phenomenon was discovered in the 1930s but was not applied to imaging until the 1970s. MRI systems became available for clinical use in the 1980s. MRI systems are the imaging system of choice for studies of the brain

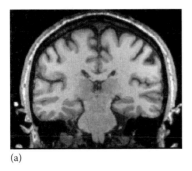
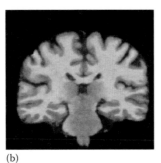
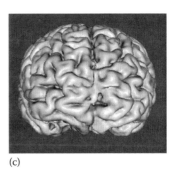

(a)　　　　　　　　　　　　(b)　　　　　　　　　　　　(c)

FIGURE 2.13 Coronal section of T1-weighted MRI of a head (a) illustrating the high soft tissue contrast possible, with little signal from the skull. It is common practice to use special processing to remove nonbrain tissues from head MRI (b). Brain surface model made from the resulting brain-only image (c) allows users to visualize the brain's surface anatomy.

and spinal cord due to their high contrast in soft tissues and lack of interference from bone (Figure 2.13).

The introduction of functional MRI (fMRI) in the mid-1990s led to its use in brain studies, previously only possible using PET. Overlays of an fMRI study onto a patient's high-resolution anatomical MRI images are often used to illustrate both the functional and anatomical spatial relationships. The developments of arterial spin labeling (ASL) for absolute blood flow and diffusion tensor imaging (DTI) for the assessment of white matter have expanded the usage of MRI for research and clinical purposes. MRI has become the most popular form of CT imaging in radiology departments supporting anatomical and functional studies for all major organ systems in the body.

HOMEWORK PROBLEMS

P2.1 The optical density difference between two regions of an x-ray film is 0.7. What is the ratio of the intensity of transmitted light between these two regions? Assuming that the film-screen has a gamma of 0.76, what is the ratio of the radiation exposure between these two regions?

P2.2 Describe the process of latent image generation for a photostimulable phosphor plate digital radiographic system. Why is the dynamic range of these image receptors much higher than that of film-screen systems?

P2.3 Estimate the reduction in scatter produced by a 1 mm collimated x-ray scanning system (like in Figure 2.9) compared with a conventional x-ray beam. Assume that the object is $100 \times 100 \times 100$ cm phantom that is placed on the image detector and that the entire phantom is imaged. Make your calculation based only on the geometrical efficiency at the center of the detector. Report the ratio of scatter using the slit to that without it.

P2.4 The relationship between pixel value and exposure for a photostimulable phosphor or flat panel detector can be determined using the following equation:

$$\text{Pixel value} = C_1 \log_{10}(E) + C_2$$

where
  E is the exposure in mR
  C_1 and C_2 are constants

You make the following measurements in an attempt to determine the relationship between pixel values and exposure:

Pixel value	1019	738	485	260	7
Exposure (mR)	100	10	1	0.1	0.01

While the exposure was measured accurately, the pixel values were a bit noisy.

(a) Use a least square error fit method to estimate the values of C_1 and C_2.

(b) Plot the raw data and the fitted data on a graph of pixel value vs. $\log_{10}(E)$.

(c) What is the R^2 value for the fit? What is the standard error in the measurement of exposure?

CHAPTER 3

Digital Imaging in Diagnostic Radiology

3.1 BASIC CONCEPTS OF DIGITAL PLANAR IMAGING

Unlike traditional plane-film radiography that generates images on film through chemical processing, digital radiography generates images using electronic processing. The transition by radiology departments to digital imaging has been gradual, often by adding digital components to older imaging technologies. An example of such a hybrid imaging system for dynamic planar imaging is illustrated in Figure 3.1. For this system, an image intensifier and a video camera acquire the x-ray images as analog signals that vary continuously with the intensity of x-rays detected. The analog video signals are converted into digital signals, having a well-defined range of discrete values. Analog-to-digital converters (ADCs) are used to encode both position and signal intensity. In full digital systems, an array of detectors determines the position within the image, and each detector's analog signal is converted to a digital value using an ADC. For both hybrid and full digital systems, the images are sent to a digital image processor that can analyze, transform, and display the image. Digital images are stored using media such as magnetic tape and magnetic, optical, or solid-state drives. The digital image is transformed back into an analog image using a digital-to-analog converter (DAC) for display on a video display terminal or on a digital monitor (LCD, OLED, etc.).

Mathematically, the digital image is a two-dimensional array of numbers and is "digital" or discrete in two respects: spatially and numerically (Figures 3.2 and 3.3).

3.1.1 Digital Position

The digital image is discrete with respect to position. If you look at a digital image with a magnifying glass, you will see it is composed of little gray squares or cells arranged in a rectangular or square array (Figure 3.2). These cells are called "pixels," an abbreviation for "picture element." The pixel is the smallest unit in a planar image and its size in

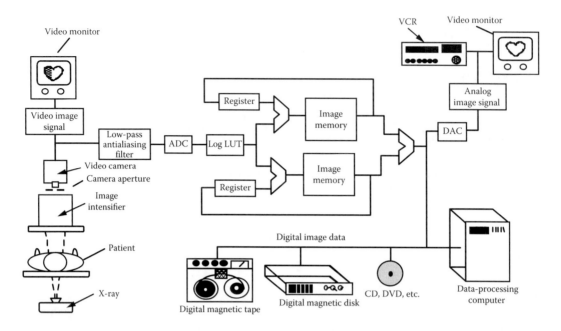

FIGURE 3.1 The components of a hybrid digital imaging system. On the left is a standard image intensifier–based fluorographic system. The ADC converts the analog video to digital images that are processed using the various registers. Display and storage are also illustrated.

medical images must be chosen carefully to retain as much detail as possible. An image with M columns and N rows of pixels is commonly referred to as an $M \times N$ (pronounced "M by N") image. A typical x-ray CT image is 512×512, although some systems produce 1024×1024 images; many investigators believe that a 2048×2048 or greater array of pixels is required for chest radiography.

3.1.2 Digital Value

The digital image is also discrete with respect to the value associated with each pixel, *pixel value*. The brightness in a digital image does not change from black to white with continuous shades of gray as it does with x-ray film (Figure 3.2). Rather, the brightness changes in small increments from black to white. Each level is represented using an integer value with the smallest representing black, the largest representing white, and intermediate values representing shades of gray, for example, these are called digital gray-scale images. Therefore, a digital image is an array of binary numbers where each number is a pixel value. As in all digital computers, these integers are processed and stored in a binary format. Figure 3.3 illustrates this using just 3 bits to store values.

Before continuing with our discussion, we will define the terms "bits," "bytes," and "kilobytes," which are commonly used in discussions of digital images and image processors (Figure 3.4). A "bit" is a "binary digit," and a binary digit can have just two values,

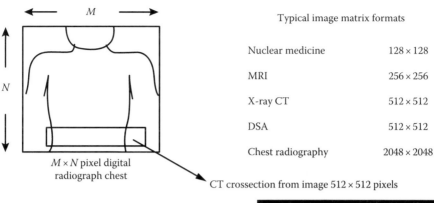

Typical image matrix formats

Nuclear medicine	128×128
MRI	256×256
X-ray CT	512×512
DSA	512×512
Chest radiography	2048×2048

$M \times N$ pixel digital radiograph chest

CT crossection from image 512×512 pixels

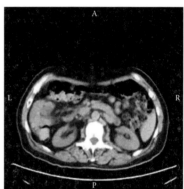

1. A digital image is discrete in both position and value.

2. A digital image is composed of pixels (picture values).

3. Each pixel value is a digital number corresponding to an image value.

4. The image value relates to some physically measurable quantity. It might be CT number, MRI signal, T1, T2, or in functional images it might have units that relate to blood flow, or some other functional measure.

FIGURE 3.2   A digital image is an array of numbers referred to as an array of picture elements or pixels. Values for each pixel are displayed for the CT section image ranging from black to white depending on pixel values.

0 and 1, just as our ordinary decimal (base 10) number system has ten values per digit, 0 through 9.

Typical medical images require 10–12 bits per pixel; a 10-bit image has 1024 (or 2^{10}) different levels of gray, while a 12-bit image has 4096 (or 2^{12}) possible different gray levels. Often it is more convenient to describe digital data in terms of "bytes," which are 8-bit groupings. You probably have heard the term "kilobyte" and assumed logically (but incorrectly) that a kilobyte was 1000 bytes. Actually, a kilobyte (abbreviated kB) is 1,024 (2^{10}) bytes, and 327,680 bytes is exactly 320 kB. Due to their large size, memory requirements for medical images usually are stated in terms of megabytes (~10^6 bytes) or gigabytes (~10^9 bytes).

3.2 DIGITAL IMAGE STORAGE REQUIREMENTS

It is important to ensure that both position and value are represented digitally with adequate precision. The number of samples across an image determines the "digital spatial precision." Likewise, the number of samples over the signal range determines the "digital

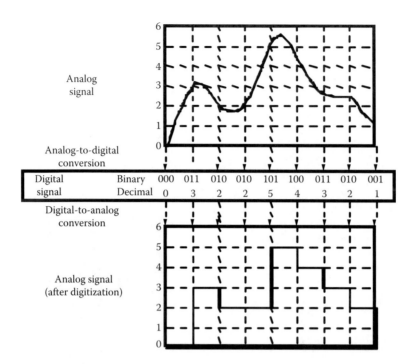

FIGURE 3.3   An example of ADC conversion using 3-bits where the output binary numbers range from 0 to 5. The upper graph is the analog signal and the lower graph the resulting digital signal. Vertical dashed lines are the points where the ADC samples the analog signal. The ADC assigned the closest digital value to the analog signal value for each sample, and the digital value is fixed until the next sample.

1 bit = 1 binary digit

1 nibble = 4 bits

1 byte = 8 bits

1 word = 2, 4, or 8 bytes

1 kilobyte (kB) = 2^{10} bytes = 1024 bytes

1 megabyte (MB) = 2^{20} bytes = 1024 kB

1 gigabyte (GB) = 2^{30} bytes = 1024 MB

1 terabyte (TB) = 2^{40} bytes = 1024 GB

FIGURE 3.4   Elementary definitions used in describing individual the digital numbers at the top and for large collections of digital numbers at the bottom (common for specifying storage capacity). Note that the most common unit is the byte (bold).

intensity precision." Spatial sampling and data storage requirements are illustrated for chest radiography in the following example:

Example 3.1

A chest radiograph is $14'' \times 17''$ in area (Figure 3.5). Assuming that we digitize the chest film with 16 bits (2 bytes) per pixel, with a pixel spacing that preserves the

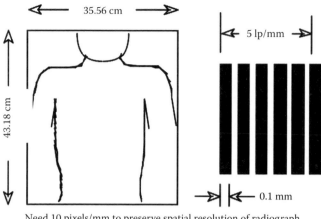

Need 10 pixels/mm to preserve spatial resolution of radiograph

FIGURE 3.5 Determining the digital spatial sampling requirement for a chest radiograph based on the sampling theorem and the limiting resolution of the radiograph.

inherent spatial resolution in the chest film (5 line pairs per millimeter or lp/mm), calculate the memory needed to store the digital chest radiograph.

Solution

A line pair consists of a dark line adjacent to a bright line and is analogous to one cycle of a square or sine wave. If we want to preserve the 5 lp/mm resolution of the chest film, we need 10 samples (or pixels) per millimeter (i.e., 5 samples of a bright pixels each next to a sample of dark pixels). This leads to 10 samples/mm and a spacing of 0.1 mm/pixel. Converting to metric units, the 14″ × 17″ chest radiograph has dimensions of 355.6 × 431.8 mm and therefore requires 3556 × 4318 pixels to preserve the fundamental spatial resolution of the radiograph. Each pixel value uses 16 bits (2^{16}= 65,536 gray shades) or 2 bytes, so the number of bytes needed to represent the chest radiograph is

$$\text{Number of bytes} = 3,556 \times 4,318 \text{ [pixels]} \times 2 \text{ [bytes/pixel]}$$
$$= 30,709,616 \text{ bytes, which is } \sim 29.3 \text{ MB}$$

Approximately 100 million chest radiographs are taken every year in the United States, virtually all of which are retained in radiology file rooms for medical–legal reasons. Storage and information retrieval difficulties can arise if chest films are stored digitally without some form of file compression.

A digital image is acquired with samples taken using uniform spacing. This supports direct encoding of pixel locations as rows and columns of a digital image array. For older analog images, an ADC is needed to encode digital positions (column–row addresses). Various means are seen for conversion from analog to digital. For example, in the digitization of a radiograph by a laser scanner, the laser and photodetector are swept across the film at a uniform speed with column–row locations encoded using digital position encoders. For nuclear medicine Anger-style gamma cameras, analog *x-y* positions are first determined, and then ADCs are used to convert these to column–row locations. Systems using an array of detectors (digital radiographic or DR panels) intrinsically encode column–row locations based on the positions of the detectors. While column–row location encoding is key to formatting digital images, we do not need to explicitly store column–row values in image files. Digital image data are generally stored in either column or row order. This order, along with data type and the number of rows and columns, is stored in the image header. When the image is loaded, the image header is read and the image is formatted into a column–row array for manipulation and display.

3.3 DIGITAL SAMPLING REQUIREMENT

3.3.1 Shannon's Sampling Theorem

It is important in digital imaging that the spatial sampling frequency (samples/distance = pixels/distance) be sufficient to preserve important information in the image, that is, the range of frequencies in the analog signal. It is important to select pixel spacing (pixels per unit distance) such that important spatial information in the analog image is not lost (highest frequency perhaps). The highest spatial frequency present in the analog signal is based on the spatial resolution of the imaging system. The importance of Shannon's sampling theorem is that it specifies the sampling frequency needed to capture this highest frequency for systems using equally spaced samples, as is the case for current medical imaging systems.

Shannon's sampling theorem: If the maximum spatial frequency in an analog signal is f_{max} (**cycles/mm**), then the signal must be digitized with sampling frequency (f_s) of at least $2f_{max}$ (**samples/mm**).

Note the difference in how we specify frequency of the analog signal as cycles/mm and the sampling frequency as samples/mm. In medical imaging, we often use line pairs/mm (lp/mm) instead of cycles/mm when referring to the analog image. The analog signal can be that of the x-ray pattern impinging on a digital image plate, an x-ray film to be scanned, or that in a hybrid system as shown in Figure 3.1.

The rationale for the $2f_{max}$ sampling requirement can be investigated by considering uniformly spaced sampling of sine waves. Assume we have a sine wave of frequency f_0 that is sampled at several sampling frequencies, ranging up to $2f_{max}$. As can be seen in Figure 3.6a, when the sampling frequency equals the frequency of the sampled sine wave (f_0), values at each sampled point are identical. The resulting sampled digital signal in this case will have a spatial frequency of zero, with a magnitude that depends on the phase of the sine wave where the sampling began. Similarly, if the function is sampled at higher frequencies where $f_0 < f_s < 2f_0$, the sampled sine wave will have a frequency greater than zero but lower than f_0. *Only when $f_s \geq 2f_0$ will the sampled f_s sine wave match the frequency of the analog sine wave.* However, oversampling (4×–10×) is needed to accurately capture the amplitude of the sine wave, a common practice with digital audio recordings.

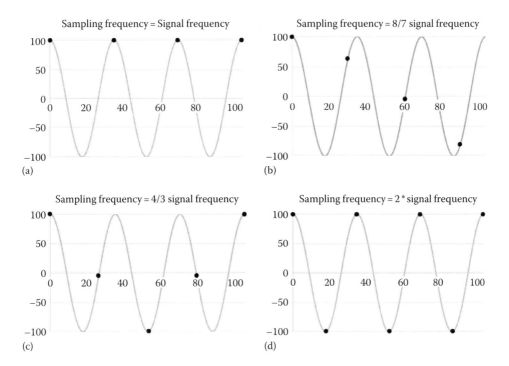

FIGURE 3.6 In digital sampling, the frequency of an analog signal is faithfully reproduced only if the sampling rate is at least twice that of the maximum signal frequency. At lower sampling rates, the approximated function will contain false signal frequencies. The introduction of these false frequencies is called "aliasing." In this example, the signal frequency is 3 cycles/mm. Sampling shown in this figure are (a) 3 samples/mm; (b) 3–3/7 samples/mm; (c) 4 samples/mm; and (d) 6 samples/mm, that is, sampling to preserve the analog signal's frequency.

3.3.2 Nyquist Frequency

The highest frequency that can be represented in sampled data is called the "Nyquist" frequency, which is exactly one-half the sampling frequency ($f_s/2$). The logic for this limit is that it takes two samples to encode both peak and valley values for each cycle of a sine wave. If this sampling requirement (two samples per cycle) is not met, then the digitized signal will contain *false* periodic signals (called *aliases*), with a frequency less than the Nyquist frequency. It will be shown (Chapter 6) that if f_s is the sampling frequency and f the frequency in the analog signal, then f will alias to a lower frequency f_{alias} when $f_s < 2f$. The aliased frequency will be determined from the following equation:

$$f_{alias} = |nf_s - f| \qquad (3.1)$$

Here, n is an integer used to ensure that $f_{alias} < f_s/2$ (i.e., less than the Nyquist frequency). For example, if an analog signal with a frequency of 60 Hz is sampled at a rate of 70 Hz (Nyquist frequency = 35 Hz), then $n = 1$, and the aliased signal will have a frequency of $|70 - 60| = 10$ Hz. In another example with $f_s = 35$ Hz (Nyquist frequency = 17.5 Hz), then $n = 2$, and the aliased signal would also have a frequency of $|2{\ast}35 - 60| = 10$ Hz, identical to the 70 Hz sampling rate. The analog signal must be sampled at twice its frequency (60 Hz × 2 =120 Hz in this example) to prevent aliasing, that is, $|120 - 60| = 60$ Hz. If $f_s > 2f$, then Equation 3.1 does not apply, and for this example, the digital sample will have the correct frequency (60 Hz). We will return to the issue of digital sampling and aliasing later, after we have developed more mathematical tools to model the effect of aliasing in digital medical images.

3.3.3 Compressed Sensing

Many images can be compressed with little (lossy) or no loss (lossless). This is especially important for PACS systems where large images must be stored and rapidly transmitted between storage devices and viewing areas. The general scheme has been to acquire images that capture detail based on the sampling theorem requirements, and use a compression scheme to reduce the data size. An alternative approach deals with reducing the amount of image data captured using a process called "compressed sensing." As opposed to classical sensing or sampling, compressed sensing uses variable spaced sampling. While not applicable to many medical imaging technologies, compressed sensing has shown promise with MRI where *a priori* knowledge of the sparse nature of signals allows varying sample spacing in the phase encode direction of k-space for compression.

HOMEWORK PROBLEMS

P3.1 The Japanese revere Beethoven's Ninth Symphony, especially Beethoven's musical rendition of Schiller's great "Ode to Joy" in the fourth movement, which celebrates the potential of humanity living together with love and peace. It is not a coincidence that the compact disc (CD) was designed by Sony so that Beethoven's Ninth Symphony would fit on a single side, providing a recording that could be played without interruption. For purposes of this analysis, we will assume that the CD stores a digital record of the amplitude waveform forming the musical sound.

(a) Technical specifications indicate that most compact disc players are capable of a signal-to-noise ratio of 96.3 dB. It is more accurate to state that the dynamic range (ratio of loudest possible sound to softest sound) is 96.3 dB, and that this dynamic range limitation is provided by the digitization of the signal. Show that this dynamic range requires 16 bits assuming that 1 bit is used to record the softest sound and 16 bits to record the loudest sound.

(b) On a CD the music is digitized at a sampling frequency of 44 kHz. Why is this sampling frequency required if the human auditory system is capable of hearing sounds only in the range of 20 Hz to 20 kHz?

(c) How many bytes does it take to record Beethoven's Ninth Symphony (approximately 1 h) on a compact disc?

(d) How much compression would be needed to completely store this symphony on a 64 MB USB RAM storage device in MP3 format? How might the quality of the sound differ when compared to the uncompressed version?

P3.2 In the plane of the detector, what is the highest spatial frequency that can be recorded by a 512 × 512 pixel digital fluoroscopy system with a 150 mm × 150 mm receptor?

P3.3 Most radiographs are obtained by placing a "grid" between the patient and image receptor. The grid is composed of thin strips of lead separated by a material (plastic, aluminum, or carbon fiber) that are transparent to x-rays. Therefore, the grid reduces scattered radiation emerging from the patient that would degrade the image, while allowing the primary photons to be transmitted to the film.

A 14″ × 17″ chest radiograph is recorded with a grid having 100 lines per inch where the width of the lead strips is equal to the width of the material between the lead strips.

(a) If the chest radiograph is digitized with a 2048 × 2048 image matrix, explain why aliasing will be produced. What will be the frequency and the appearance of the aliased signal?

(b) What digital sampling spatial frequency would avoid aliasing? What is the digital image format and the pixel size (in inches) that corresponds to this digital sampling frequency?

P3.4 In digital subtraction angiography, an analog video signal is converted to a 10-bit digitized signal. However, the video signal can have a bandwidth, which exceeds 7.5 MHz, requiring the use of a low-pass "antialiasing filter" to condition the signal before it is digitized to a 512 × 512 format with an ADC. Assume that the video frame (one image) consists of 525 lines, with a frame time of 1/30th s, and that each line of the frame is digitized with 512 samples.

(a) If no antialiasing filter is used, calculate the temporal frequency of the aliased signal produced in the digitized signal.

(b) Calculate the bandwidth of the low-pass filter that should be used to prevent aliasing in the digital image.

II

Intermediate Concepts

CHAPTER 4

Physical Determinants of Contrast

A MEDICAL IMAGE CAN BE roughly described in terms of the three basic features given in Chapter 1: contrast, spatial resolution, and noise. Spatial resolution or clarity refers to the spatial detail of small objects within the image. Noise refers to the precision within the image; a noisy image will have large fluctuations in the signal across a uniform object, while a precise signal will have very small fluctuations. The subject of this chapter, contrast, relates to the difference in signals between a structure and its immediate surroundings. For example, if circles are displayed against a black background, a white circle will have larger contrast relative to the background when compared to gray circles (Figure 4.1). One uses the differences in gray shades to "visually" distinguish different tissue types, determine anatomical relationships, and sometimes assess their physiological functions. The larger the contrast between different tissue types, the easier it is to make distinctions clinically. It is often the objective of an imaging system to maximize the contrast in the image for a particular object or tissue of interest, although this is not always true since there may be design compromises where noise and spatial resolution are also very important. The contrast in an x-ray image depends on both physical characteristics of the object and properties of the device(s) used to image the object. The focus of this chapter is x-ray image contrast and we discuss the physical determinants of contrast, including material properties, x-ray spectra, detector response, and the role of perturbations such as scatter radiation and image intensifier veiling glare. We include physical determinants of contrast for several other medical imaging modalities, including those used in nuclear medicine, magnetic resonance imaging, and computed tomography to round off the discussion.

4.1 COMPONENTS OF X-RAY IMAGE CONTRAST

Contrast can be quantified as the fractional difference in a measurable quantity between adjacent regions of an image. Usually, when we say "contrast," we mean image contrast, which is the fractional difference in signals between two adjacent regions of an image.

| Low contrast | Medium contrast | High contrast |

FIGURE 4.1 Contrast relates to the relative difference between an object (circle here) and its surrounding background. Here the background gray level is set to black and the object gray level increased from a low-contrast image on the left to a high-contrast image on the right.

TABLE 4.1 Sources of Image Contrast

Radiographic Contrast
 Material thickness
 Physical density
 Electron density
 Elemental composition (effective Z)
 X-ray photon energy
Detector Contrast
 Detector type (film vs. electronic)
 Film characteristic curve (H&D)
 Spatial response of detector
Display Contrast
 Window and level settings
Physical perturbations
 Scatter radiation
 Image intensifier veiling glare
 Base and fog of film

In conventional radiography, contrast can be separated into three components: (1) radiographic contrast, (2) detector contrast, and (3) display contrast (Table 4.1 and Figure 4.2).

4.1.1 Radiographic Contrast

Radiographic contrast (sometimes called subject contrast) refers to the difference in x-ray photon fluence emerging from adjacent tissue regions in the object. Radiographic contrast depends on differences in atomic number, physical density, electron density, thickness, and the energy spectrum of the x-ray beam. Because radiographic contrast depends on both subject and nonsubject factors (x-ray energy), we avoid the use of the term "subject contrast" though it is commonly seen in many publications. Radiographic contrast is the basis of overall contrast, that is, without radiographic contrast, the other components can have no effect on the overall contrast.

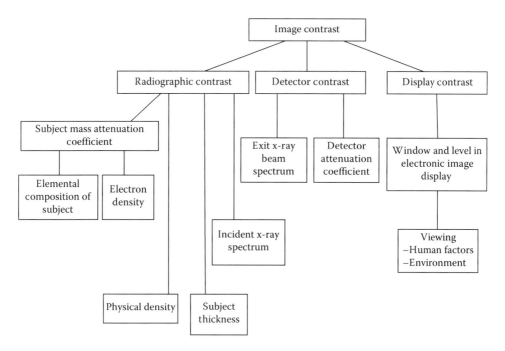

FIGURE 4.2 Image contrast is broadly divided into three components: radiographic contrast, detector contrast, and display contrast. There are distinct subcomponents of each.

4.1.2 Detector Contrast

All detectors alter radiographic contrast. As we shall see later, for linear system analysis, the measurement of the detector's image signal must be linearly related to the intensity of the radiation forming the image for analyses. The H&D curve is used to convert film densities to relative exposure values. Electronic detectors provide a means to convert signals to absolute exposure values. Detector contrast depends on the chemical composition of the detector material, its thickness, atomic number, electron density, as well as the physical process by which the detector converts the radiation signal into an optical, photographic, or electronic signal. Detector contrast also depends on the x-ray spectrum exiting the object. *The detector may increase or decrease contrast relative to the radiographic contrast,* that is, the detector may produce signals that have a larger or smaller fractional difference between adjacent areas of the image in comparison with the difference in exposure.

4.1.3 Display Contrast

An observer can significantly alter the displayed contrast by adjusting monitor's display settings to increase, decrease, or preserve contrast.

4.1.4 Physical Perturbations

X-ray image contrast can also be affected by various perturbations such as scattered radiation, image intensifier veiling glare, and the base and fog density levels of film. These perturbations reduce the image contrast.

4.1.5 Other Confounds

Finally, training, physiological status of the human viewer (e.g., myopia, astigmatism, color blindness, fatigue) as well as environmental factors such as ambient light levels in the viewing area also affect how one perceives the contrast presented in the image. In this book, we will not include these factors with the other components of contrast. Nevertheless, they are essential to the diagnostic task and deserve mention.

4.2 IMAGE CONTRAST MATH

The physical determinants of x-ray image contrast can be understood by examining the processes by which a radiographic image is formed. We will consider a system in which a patient is placed between an x-ray tube and a detector: The detector is usually either a film screen or an electronic screen. The x-ray tube is operated at a certain peak kilovoltage (kVp), which, along with filtration, determines the energy spectrum of beam. X-ray photons from the source are attenuated by various materials in the patent (muscle, fat, bone, air, and contrast agents) along the path between the source and the detector where the image is formed.

The x-ray attenuation of each tissue depends on its elemental composition as well as the energy of the x-ray photons. Attenuation is modeled using a linear attenuation coefficient (μ), which is the fraction of photons absorbed by a unit thickness of the tissue.

Equation 4.1 relates the photon fluence entering (Φ_0) to that exiting (Φ) a volume of tissue. This equation uses the mass attenuation coefficient (μ_m) rather than the linear attenuation coefficient (μ) since it is more commonly tabulated, and there is a simple relationship between these $\mu_m(E) = \mu(E)/\rho$, where ρ is the density of the attenuator:

$$\Phi(E,x) = \Phi_0(E)e^{-\mu(E)x} = \Phi_0(E)e^{-(\mu(E)/\rho)\rho x} = \Phi_0(E)e^{-\mu_m(E)\rho x}, \quad (4.1)$$

where

$\Phi_0(E)$ is the photon fluence entering the volume of homogeneous material at energy E
$\Phi(E, x)$ is the photon fluence exiting the volume of thickness x
$\mu_m(E)$ is the mass attenuation coefficient of the material at energy E
ρ is the density of the material

The tissue-specific factors that affect radiographic contrast are revealed in the three parameters in the exponent of Equation 4.1. The first parameter is the mass attenuation coefficient (μ_m), which depends explicitly on energy (E) and implicitly on the atomic number of the material and its electron density. The second parameter is the mass density (ρ) of the material. The third parameter is the thickness of the material (x). An increase in any of the three parameters increases attenuation. However, in x-ray imaging, only energy is readily adjustable, and changing the x-ray machine's kVp setting does this.

If we measure photon fluence at a single energy on the detector-side of the patient, we can determine radiographic contrast using Equation 4.2. For example, if behind the patient, a photon fluence of Φ_1 is measured in the surroundings and a photon fluence of

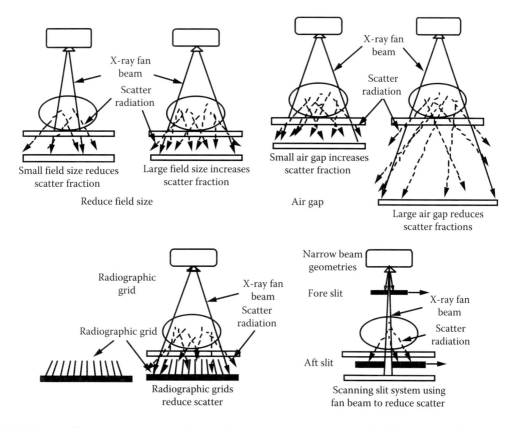

FIGURE 4.3 Planar x-ray imaging done with an x-ray source on one side of the patient's body and a planar detector on the other side. X-ray photons can pass through the patient's body without interacting (primary photons), be scattered (secondary photons), or completely absorbed. Secondary photons reduce contrast, so various methods are illustrated to reduce their impact.

Φ_2 is measured behind the object of interest (Figure 4.3), then the radiographic contrast would be

$$C = \frac{\Delta\Phi}{\Phi_1} = \frac{\Phi_2 - \Phi_1}{\Phi_1}. \tag{4.2}$$

For a radiopaque object, one where $\Phi_2 = 0$, the radiographic contrast $C = -1$. Here radiographic contrast would be reported as negative 100%. When $\Phi_1 < \Phi_2$, radiographic contrast is positive and can exceed 100%. If $\Phi_1 = \Phi_2$, the object cannot be differentiated from its surroundings since radiographic contrast is zero ($C = 0$). An x-ray beam is emitted with a spectrum of energies (Figure 4.6), complicating the use of Equation 4.1 to calculate radiographic contrast; however, Equation 4.2 is helpful when modeling an x-ray beam using a single "effective" energy.

The magnitude of mass attenuation coefficients decreases with increasing energy, with attenuation also decreasing except at points of discontinuity called absorption edges, mostly the K-edge or L-edge (Figure 4.4). At these energies, photoelectric interactions cause large increases in absorption as the photon energy slightly exceeds the binding energy of

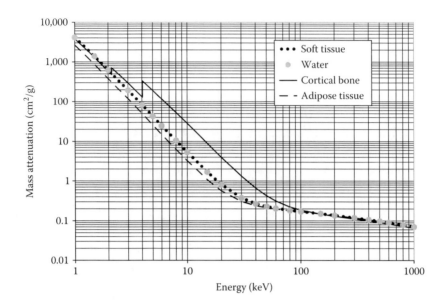

FIGURE 4.4 The mass attenuation coefficients for tissues of interest and water (tissue equivalent model) decrease rapidly with increasing energy in the diagnostic x-ray imaging range (10–100 keV). The exception is at absorption edges as illustrated for cortical bone below 10 keV.

orbital electrons in the K and L shells. Overall, contrast tends to decrease as photon energy increases except when K-edge or L-edge discontinuities cause large increases for the tissue of interest. (As we will see, this is a consideration when high-Z contrast agents are utilized.)

4.3 RADIOGRAPHIC CONTRAST OF BIOLOGICAL TISSUES

Radiographic contrast due to differences in attenuation in the body depends somewhat on the difference in mass attenuation coefficients of the various tissues. This intrinsic radiation attenuation property for each tissue type is determined by its elemental and chemical composition, whether a compound or mixture. To simplify dealing with the complex nature and variety of tissues, we will consider the body as being composed of three main tissues: fat (adipose tissue), soft tissue, and bone. In addition to these tissues, we will consider attenuation properties of air found in the lungs (and in the gastrointestinal tract) as well as the properties of contrast agents when introduced into the body.

The elemental compositions of the main tissue types are given in Table 4.2. Physical properties relevant to attenuation are given in Table 4.3, where they are ordered in the ascending order, from lowest to highest attenuation. Water is included in these tables because its attenuation properties are similar to those of soft tissues. Corresponding tissue mass attenuation coefficients are plotted in Figure 4.4 as a function of energy.

4.3.1 Soft Tissue

The term "soft tissue," as used in this text, excludes fat but includes muscle and body fluids. The term lean soft tissue is sometimes used to describe nonfatty soft tissues, but we will use the less cumbersome term soft tissue and implicitly exclude fat. There are, of course,

TABLE 4.2   Elemental Composition of Body Materials Ordered by Increasing Atomic Number (Z)

% Composition (by Mass)	Fat (Adipose Tissue)	Soft Tissue (Striated Muscle)	Water	Bone (Femur)
Hydrogen (**Low Z =1**)	**11.2**	**10.2**	**11.2**	**8.4**
Carbon ($Z = 6$)	**57.3**	12.3		**27.6**
Nitrogen ($Z = 7$)	1.1	3.5		2.7
Oxygen ($Z = 8$)	**30.3**	**72.9**	**88.8**	**41.0**
Sodium ($Z = 11$)		0.08		
Magnesium ($Z = 12$)		0.02		0.2
Phosphorus ($Z = 15$)		0.2		7.0
Sulfur ($Z = 16$)	0.06	0.5		0.2
Potassium ($Z = 19$)		0.3		
Calcium (**High Z = 20**)		0.007		**14.7**

TABLE 4.3   Physical Properties of Human Body Constituents

Material	Effective Atomic No.	Density (g/cm³)	Electron Density (electrons/kg)	
Air	7.6	**0.00129**	3.01×10^{26}	Low attenuation
Water	7.4	1.00	3.34×10^{26}	
Soft tissue	7.4	1.05	3.36×10^{26}	
Fat	**5.9–6.3**	**0.91**	3.34–3.48×10^{26}	
Bone	**11.6–13.8**	**1.65–1.85**	3.00–3.19×10^{26}	High attenuation

many different types of soft tissues including liver tissue, collagen, ligaments, blood, cerebrospinal fluid, and so on. However, the chemical composition of all soft tissues is dominated by elements with low atomic numbers (Table 4.2). We will assume that they are radiographically similar to water and have an effective atomic number of 7.4 and an electron density of 3.34×10^{26} electrons per gram. This assumption is plausible since, as Ter Pogossian has pointed out; soft tissues are approximately 75% water by weight, while body fluids are 85% to nearly 100% water by weight.

Water-equivalent soft tissues have several important radiologic properties that contribute to their contrast. First, the photoelectric effect dominates photon attenuation up to energy of ~30 keV, the region delimited by the steeper slopes of the curves in Figure 4.4. The Compton effect becomes increasingly dominant in soft tissues over the remainder of the diagnostic energy range. Photoelectric interactions provide better differentiation between tissue types due to the larger differences in mass attenuation coefficients. Therefore, it is desirable to use lower-energy photons to maximize contrast in diagnostic examinations of soft tissues. Second, because the body is about 70% water-equivalent by weight, contrast for these tissues is dictated predominantly by variations in thickness or density (i.e., by the ρx product in Equation 4.1). In the diagnostic energy range, the half-value layer (HVL) of soft tissue is 3–4 cm so that a thickness difference of 3 cm is needed to provide a radiologic contrast of 50%. Finally, the radiographic similarity of a majority of soft tissues in the human body complicates our imaging task. For example, it is impossible to visualize blood directly or to separate tumors from surrounding normal soft tissue without special procedures. This forces radiologists to use "contrast agents" (discussed in the following) to provide contrast and enable visualization of these structures. Without contrast agents, except in the grossest manner, it is impossible to visualize structural details of important structures such as the liver, GI tract, or cardiac blood pool, using standard radiographic imaging techniques.

4.3.2 Fat

The energy of the x-ray beam, electron density, physical density, and atomic number determine the attenuation capability of any material. Due to its many low atomic number elements, fat has a lower physical density and lower effective atomic number (Tables 4.2 and 4.3) than either soft tissue or bone, and therefore, has a lower photoelectric attenuation coefficient. For this reason, fat has a lower total attenuation coefficient than other materials in the body (except air) at low x-ray energies where photoelectric interactions are the dominant effect.

Unlike other elements, the nucleus of hydrogen is free of neutrons, giving hydrogen a higher electron density (electrons/mass) than other elements. Because hydrogen contributes a larger proportion of the mass in fat than it does in soft tissue and bone, fat has a higher electron density than other tissues. This becomes particularly important at higher energies where Compton interactions dominate attenuation (interactions with loosely bound electrons). In fact, inspection of the graph of mass attenuation coefficients (Figure 4.4) shows that at higher energies (>100 keV), the mass attenuation coefficient of fat slightly exceeds that of bone or soft tissue, due to the higher electron density of fat. However, due to its low density, it does not have a higher linear attenuation coefficient.

As shown in Tables 4.2 and 4.3, the differences in atomic number, physical density, and electron density between soft tissue and fat are slight. The differences in the linear attenuation coefficients and therefore in radiographic contrast between fat and soft tissue can be small. One must use the energy dependence of the photoelectric effect to produce radiographic contrast between these two materials. This is particularly true in mammography where one uses a low-energy x-ray beam (effective energy ~18 keV). Such a low-energy spectrum increases the contrast between glandular tissue, connective tissue, skin, and fat, all of which have similar attenuation coefficients at higher x-ray energies.

4.3.3 Bone

The mineral content of bone results in high contrast relative to other tissues for x-ray energies the diagnostic range. This is due to two properties listed in Table 4.3. First, its physical density is 60%–80% higher than soft tissues. This increases the linear attenuation coefficient of bone by a proportionate fraction over that of soft tissue, greatly increasing attenuation. Second, its effective atomic number (about 11.6) is significantly higher than that of soft tissue (about 7.4). The photoelectric mass attenuation coefficient varies approximately with the cube of the atomic number, so the photoelectric mass attenuation coefficient for bone is about $[11.6/7.4]^3 = 3.85$ times that of soft tissue. The combined effects of greater physical density and larger effective atomic number give bone a photoelectric linear attenuation coefficient much greater than that of soft tissue or fat. This difference decreases at higher energies, where the Compton effect becomes more dominant. However, even at higher energies, the higher density of bone still allows it to have good contrast with respect to both soft tissue and fat. Therefore, when imaging bone, one can resort to higher energies to minimize patient exposure while maintaining reasonable contrast instead of resorting to low x-ray beam energies as is needed to differentiate fat from soft tissue.

4.3.4 Contrast Agents

Most of the methods we use to improve contrast involve control of variables external to the patient, such as detector response, display controls, and choice of x-ray tube kVp. The factors that control contrast within the patient, such as the thickness, physical density, and elemental composition of the body's tissues, are difficult if not impossible to control while an image is being recorded.

There are times, however, when the composition of the body part can be modified to alter radiographic contrast. This is accomplished by introducing a material, called a contrast agent, into the body to increase, or sometimes decrease, the attenuation (Table 4.4). For example, contrast agents containing iodine are commonly injected into the circulatory system when an angiographer is imaging blood vessels. This is necessary because blood and soft tissue have attenuation coefficients essentially equal to that of water. Therefore, the blood cannot be distinguished from surrounding soft tissue structures using conventional x-ray techniques without contrast agents. When an iodinated contrast agent is introduced into the circulatory system, it increases the x-ray attenuation of the blood, allowing the radiologist to visualize the blood in the vessels (either arteries or veins) or in the cardiac

TABLE 4.4   Examples of Contrast Agents

Hydopaque (Iodine)

Composition: 0.25 g $C_{18}H_{26}I_3O_9$ + 0.50 g $C_{11}H_3I_3N_2O_4$ + 0.6 g water

Physical density: 1.35 g/cm³

K-edge energy: 33.2 keV

Atomic number of iodine: 53

Applications: angiography, genitourinary (GU) studies

Barium sulfate (BaSO₄)

Composition: 450 g barium sulfate + 25 mL water

Physical density: 1.20 g/cm³

K-edge energy: 37.4 keV

Atomic number of barium: 56

Applications: gastrointestinal (GI) studies

Air

Composition: 78% N_2 + 21% O_2

Physical density: 0.0013 g/cm³

K-edge energy: 0.4 keV

Effective atomic number: 7.4

Applications: GI studies, pneumoencephalography

chambers. Barium is another element that is commonly used in contrast agents, particularly in the gastrointestinal tract. A thick solution containing barium is introduced into the gastrointestinal tract by swallowing or through some other alternate paths. When the barium solution is inside the GI tract, the walls of the tract can be visualized so that the radiologist can look for ulcerations or ruptures.

Iodine and barium are used as contrast agents for several reasons. First, they can be incorporated into chemicals that are not toxic even in large quantities; these are then introduced into the body. Second, to be useful as a contrast agent, the material must have an attenuation coefficient that is different from that of other soft tissues in the human body (Figure 4.5). When iodinated contrast agents are used in angiography, the iodine must provide sufficient x-ray attenuation to provide discernable contrast from surrounding soft tissues (represented by water in Figure 4.5).

Both barium ($Z = 56$) and iodine ($Z = 53$) meet these requirements. An older common contrast agent is Hydopaque (Figure 4.4), a cubic centimeter (cm^3) of which contains $C_{18}H_{26}I_3O_9$, $C_{11}H_3I_3N_2O_4$, and water with a density greater than that of water (1.35 g/cm^3). Most of its attenuation is provided by the iodine component due to its higher atomic number and the increased physical density of the solution. Importantly, the K-edge of iodine is at 33.2 keV near the peak of the energy spectrum used for many diagnostic studies (Figure 4.5). Similarly, the barium contrast agent used in abdominal studies contains 450 g of barium sulfate ($BaSO_4$) in 25 mL of water; this gives a suspension with a physical density of 1.20 g/cm^3. The K-edge of barium occurs at 37.4 keV, again lying near the peak of the energy spectrum used for abdominal studies.

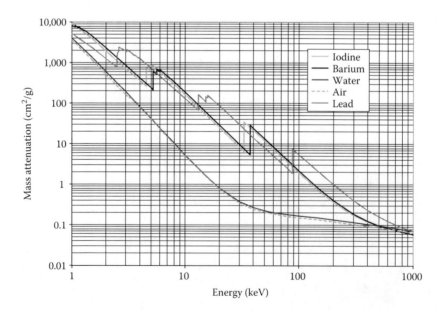

FIGURE 4.5 **(See color insert.)** Both iodine and barium and K-absorption edges in the range of diagnostic x-ray imaging. Their mass attenuation coefficients are significantly larger than those of water (tissue equivalent) and air, so they are excellent contrast agents to distinguish vessels or gastrointestinal tract bounded by air and/or soft tissues.

One can maximize the contrast with iodine and barium contrast agents by imaging with x-ray photons just above the K-edge of these elements. This can be achieved by "shaping" the x-ray spectrum adjusting kVp and using an additional metallic filter with a K-edge higher than the K-edge of the contrast agents. In Figure 4.6a, note that the 120 kVp x-ray beam contains many high-energy x-rays where iodine has a low attenuation coefficient. Also, the 60 kVp spectrum contains many lower-energy photons, below the iodine K-edge (Figure 4.6b). Adding a filter with K-edge higher than the K-edge of the contrast agent leads to a large proportion of the transmitted photons just above

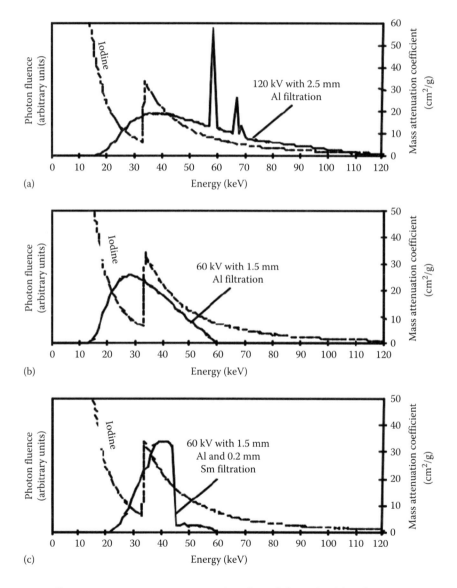

FIGURE 4.6 The x-ray spectrum varies with selected kVp (a, b). The attenuation can be optimized for iodine contrast by adding a samarium filter to the inherent aluminum filter of an x-ray system (c).

the K-edge of the contrast agent (Figure 4.6c). *Since this photon energy range falls in a region of maximum attenuation by the contrast agent, it will improve its contrast producing ability.* Metals with slightly higher atomic numbers than the contrast agents are useful as x-ray beam filters in such applications since they also have slightly higher K-edges. Rare-earth metals such as samarium (Sm) and cerium (Ce) are commonly used to filter the x-ray beam in contrast studies involving iodine or barium. Because of the principle of their operation, they are called *K-edge* filters. A rare-earth filter (e.g., samarium; K-edge = 46.8 keV) removes lower- and higher-energy photons and shapes the energy spectrum for improved absorption by an iodine-based contrast agent (Figure 4.6c).

A third contrast agent, one that reduces rather than increases x-ray attenuation, is air. Prior to the advent of computed tomography and MRI, contrast in radiographic images of the head was obtained after injecting air into the cerebral ventricles through a catheter inserted into the spinal column. The introduction of air displaced cerebral spinal fluid (CSF) in the ventricles, which otherwise have essentially the same x-ray attenuation properties as the surrounding brain tissue. Without the introduction of a contrast agent, the ventricles could not be visualized with standard x-ray techniques, nor could the various white and gray matter structures. The air in the ventricles displaced the CSF, allowing the radiologist to visualize the shape of the ventricles. Distortion in shape suggested the presence of a tumor or other abnormality. X-ray CT and MRI have practically eliminated the use of this technique, called pneumoencephalography. X-ray CT and MRI provided phenomenal advances for studies of the brain, particularly since pneumoencephalography was quite painful and prone to complications; in rare instances, it may lead to death. However, it remains an example of how air can be used as a contrast agent to visualize structures that normally would be invisible in a radiograph.

Another study where air is used as a contrast is imaging in the gastrointestinal (GI) tract. In the so-called double contrast study, a barium contrast agent is introduced followed by injecting air into the GI tract to displace the bulk of the barium contrast agent. This approach leaves behind a thin coating of barium on the inner wall of the GI tract. This allows the radiologist to observe the intricate structure of the GI lining to look for ulcerations. The property of air that makes it a useful contrast agent is its low physical density. The linear attenuation coefficient of a material is directly proportional to its physical density, and air has a linear attenuation coefficient several orders of magnitude lower than any other body tissues (Figure 4.7). However, it is not possible to introduce air into most other parts of the body.

4.4 DETECTOR CONTRAST

In the previous section, we presented methods by which radiographic or subject contrast could be maximized. However, a human observer never directly responds to the radiographic contrast since we are not equipped with biological sensors that can detect photons with energies in the diagnostic range. Instead the x-ray fluence pattern is converted to an intermediate usable form by a film-screen or electronic detector. Image contrast depends on the choice of the detector.

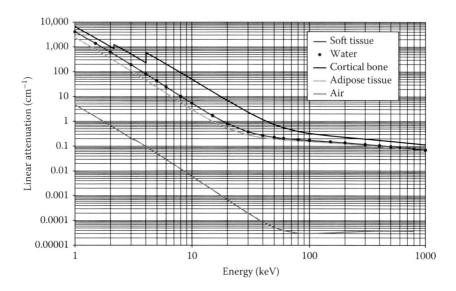

FIGURE 4.7 **(See color insert.)** The linear attenuation coefficient is the product of the mass attenuation coefficient (cm²/g) and density (g/cm³). This figure shows the tremendous difference between air and other body tissues, making it an excellent contrast agent.

4.4.1 Film-Screen Systems

An x-ray beam transmitted through a patient enters a film-screen cassette and strikes the intensifying screen, which gives off light to expose the film's photographic emulsion. The amount of light from the intensifying screen depends on the energy fluence (energy/area) of the photons as well as the chemical composition of the intensifying screen and its thickness. The chemical composition/thickness of the intensifying screen determines how much radiation is absorbed and how much light is given off per unit of absorbed radiation.

Since different regions of the body contain tissues with different elemental compositions, the energy spectrum of the beam emerging from the body and striking the image receptor can vary from one region of the body to another. In a few cases, this can contribute to differences in contrast. For example, if one region of the body contains an iodinated contrast agent, then x-ray photons with energy below the iodine K-edge will present a small radiographic (or subject) contrast, while those just above the iodine K-edge will present a much larger radiographic (or subject) contrast. A screen that is more sensitive to the photons above the K-edge, therefore, will generate an image with a higher contrast than a screen that is more sensitive to photons below the K-edge. In most cases, this is not an important effect for film-screen radiography. Therefore, one typically chooses the screen to be most sensitive to the entire spectrum of the beam emerging from the patient to minimize image noise rather than choosing the screen to be sensitive to one particular part of the energy spectrum to maximize contrast. Screens are selected based on their speed (~1/sensitivity) and resolution (to be covered in a later chapter).

In comparison with the intensifying screen in a film-screen system, the film can have a dramatic effect on image contrast. A screen responds linearly by emitting light proportional to the x-ray exposure over four to five decades of exposure. However, photographic film does not respond linearly to the light emitted by the screen. Rather, the characteristic (or H&D) curve relating film exposure to film density is sigmoid ("S-shaped") (Figures 4.8 through 4.10). A certain level of exposure has to be reached before the optical density of the film begins to change linearly with exposure. Beyond this linear range, which is one to two decades, additional exposure of the film results in a smaller increase in its optical density. The photographic film, therefore, has both a "toe" (at low exposure levels) and "shoulder" (at high exposure levels) where contrast is low compared to the linear range (Figure 4.8). Therefore, film exposure must be chosen carefully so that contrast between tissues of interest falls within the film's linear region (Figure 4.9).

An important determinant of contrast with a film-screen system is the film-gamma or film-gradient, which is the slope of the H&D curve in its linear region. If presented with the same radiographic contrast, a film with a larger gamma will produce a higher contrast image than one with a smaller gamma (Figure 4.10). The disadvantage of film with a large gamma is usually diminished latitude. Therefore, studies such as chest radiography, with a large difference in photon fluence between the lung and the mediastinum or abdomen, often utilize films with higher latitude and accordingly with lower contrast. Alternatively, studies such as mammography use high-contrast films since the images are obtained with compression of the breast to reduce unwanted exposure latitude. *When a film-screen system is used, the characteristic curve of the film is the most important component in detector contrast.*

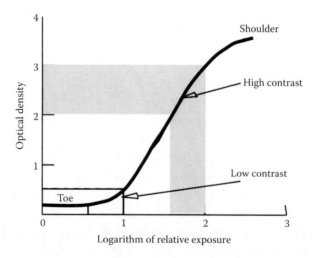

FIGURE 4.8 The characteristic sigmoidal curve of film density vs. relative exposure is called an H&D curve. Film contrast is related to the slope of this curve, so the "toe" and "shoulder" regions lead to low contrast. Highest contrast is, therefore, between the toe and shoulder regions of exposure.

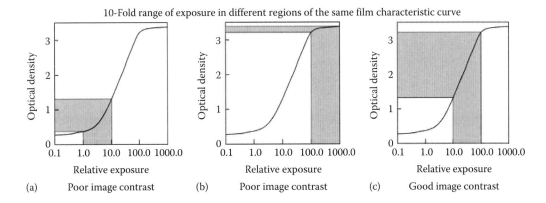

10-Fold range of exposure in different regions of the same film characteristic curve

(a) Poor image contrast (b) Poor image contrast (c) Good image contrast

FIGURE 4.9   Examples of exposures in the "toe" region of the H&D curve (a), the "shoulder" region (b), and the region between (c). For the same range of exposure, the largest range in resulting film density is in the midrange where contrast is highest.

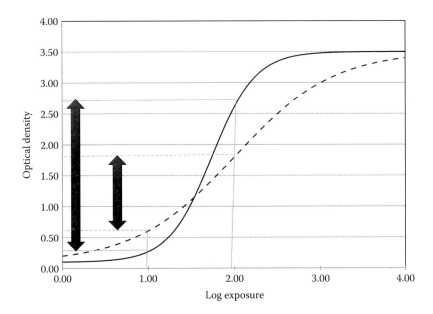

FIGURE 4.10   Simulated H&D curves for a high-contrast film (solid line) and a high latitude film (dashed). Vertical gray lines show the upper lower exposure limits within an image. The horizontal lines show the resulting upper and lower OD values for each film. The vertical arrows highlight the higher contrast capability of the high-contrast film.

4.4.2 Electronic Detectors

There are numerous materials used in electronic detectors including sodium iodide, bismuth germinate, calcium tungstate, cadmium tungstate, calcium fluoride, cesium iodide, ultrapure germanium, xenon gas, and arrays of high-Z solid-state detectors. As in the case of intensifying screens, each of these materials has an individualized spectral response. Depending on its composition and thickness, each material will absorb x-ray photons as a function of energy to various degrees. In the case of a scintillation detector, the light output

Φ from the scintillator depends on the total amount of energy absorbed by the detector material. We can calculate this value using the following equation:

$$\Phi = K \int S(E)\left\{1 - e^{-\mu_m(E)\cdot\rho x}\right\} E dE, \tag{4.3}$$

where

    $S(E)$ is the fraction of x-ray energy spectrum between E and $E + dE$
    x is the thickness of the detector
    K is the fraction of absorbed x-ray energy that is converted to light energy

Equation 4.3 can be used with any electronic detector if K is known for the detector. Regardless of the active detector material, electronic detectors differ from film-screen detectors, because unlike film their response is linear across a wide range of radiation exposure. While film-screen detectors generally have a linear response to radiation exposure that spans 1–2 orders of magnitude (Figure 4.10), an electronic detector will often have a linear response spanning 4–5 orders (Figure 4.11). As discussed in the previous section, if a radiographic image has a wide range of exposure, some exposure will occur outside of the linear range of the film characteristic curve (Figure 4.9). Exposures outside this linear

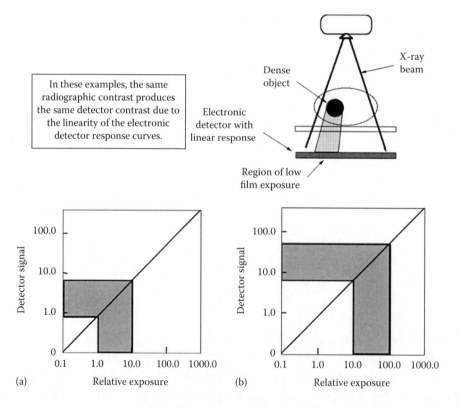

FIGURE 4.11 Unlike film, an electronic detector can respond linearly over multiple orders of magnitude, so whether the exposure is low (a) or high (b), the full range is recorded linearly.

range reduce contrast, which is an inherent limitation of film. In comparison, *the wide dynamic range of the electronic detector maintains the linearity in the radiographic signal,* even when imaging an object that has a large range in x-ray attenuation. This property is touted as an important advantage of electronic detectors.

4.5 DISPLAY CONTRAST

Digital images support the use of postacquisition enhancement to alter image contrast in a way that is not available with film-screen systems. Digital images are stored as arrays of numbers (pixels with pixel values for simplicity) rather than as optical density with film/ screen radiographs. With digital image display systems, the translation from a pixel value to its displayed brightness is done using a "look-up table" (LUT), which specifies an image brightness for each pixel value (Figure 4.12).

The operator controls display contrast by setting the range of image values to display (Figure 4.13). This range is mapped from black (minimum) to white (maximum). A wide range of pixel values results in low display contrast (Figure 4.13a). Alternatively, a narrow range of pixel values provides high display contrast (Figure 4.13b, c). A narrow range can be centered over different image values so that structures of low or high values can be displayed with high contrast. The LUT can be switched from displaying images in a positive sense where high pixel values are white (Figure 4.13d) to a negative sense where low pixel values are white (Figure 4.13e).

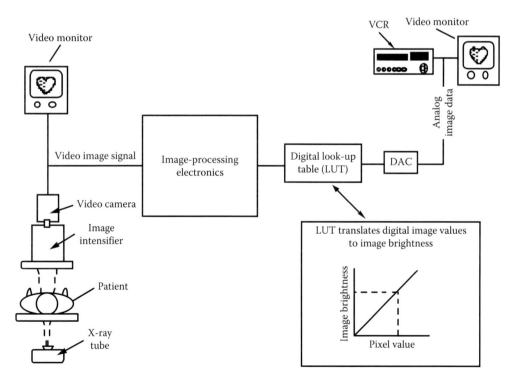

FIGURE 4.12 The display of digital images based on mapping the image's pixel values to values used by the display system. This is accomplished using an LUT where the pixel value is used to index (look up) the value for the display system.

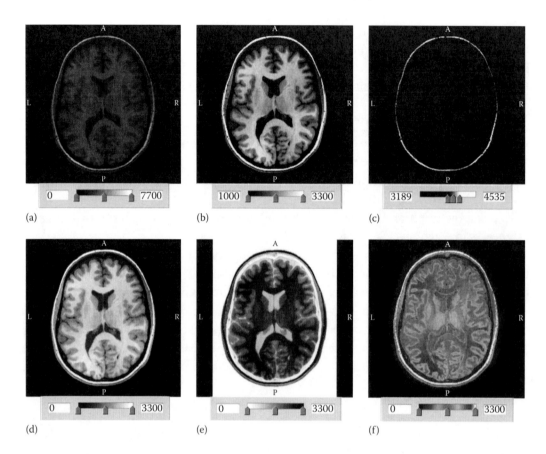

FIGURE 4.13 **(See color insert.)** MRI brain image with display range settings (a) over the full range of values (0–7700), (b) for high contrast between gray and white matter (1000–3300), (c) highlighting fat tissue in the scalp (3189–4535), (d) spanning brain tissue values (0–3300), (e) range as in (d) but with a negative LUT, and (f) a color LUT can be used to distinguish tissue ranges by color.

Digital controls provide image display flexibility not available with conventional film-screen systems including display using a color LUT. Such colors are referred to as "pseudocolors," since unlike light photographic images medical images do not represent color. In most planar x-ray imaging applications, pseudocolor is not used because transitions from one hue to the next can create a false impression of borders. However, color displays can be helpful in parametric or functional imaging where pixel values represent blood flow, transit time, or some other physiological measurement derived from images. Also, color can help to compare regions of the same value in distant parts of an image where gray-scale comparison is difficult. Color is often used in tomographic image studies, especially in radionuclide imaging (SPECT and PET) to help visualize radiotracer levels. Color is also used to overlay results from functional imaging (fMRI, PET, SPECT) onto an individual's anatomical image (CT or MRI). Also, color rendering can be used with surface model displays derived from CT and MR images in an attempt to provide a more realistic appearance or to show deep structures hidden behind superficial ones. However, it is important to remember that these electronically generated color images should not be confused with the normal concept of real world colors, rather they are pseudocolor images.

4.6 PHYSICAL PERTURBATIONS

There are three related physical perturbations that act to decrease contrast in radiographic images, scattered radiation, fog in the film, and veiling glare in the image intensifier. Fog in film can be minimized by proper storage and handling, but otherwise cannot be avoided and will not be discussed further. We will present scatter radiation and briefly discuss the similar effects of veiling glare.

4.6.1 Scattered Radiation

Scattered radiation is present in all radiographic studies and is an important consideration especially in diagnostic examinations that use broad area beams, instead of narrow (fan or pencil) beam geometry. Measurements by Sorensen show that, in the standard chest radiograph, scatter radiation can account for 50% of the radiographic signal behind the lungs and up to 90% of the signal behind the mediastinum and diaphragm when no scatter rejection techniques are used.

The image formation process in diagnostic radiology essentially captures the radiographic "shadow" cast by the beam of x-rays emitted from the focal spot within the x-ray tube. This assumes that the direction of x-ray photons is not altered when interacting within the body, and such x-ray photons are called *primary photons*. X-ray photons that scatter within the body are called secondary or *scattered photons* and appear to arise from the position in the body where the scattering occurred rather than the focal spot (Figure 4.14).

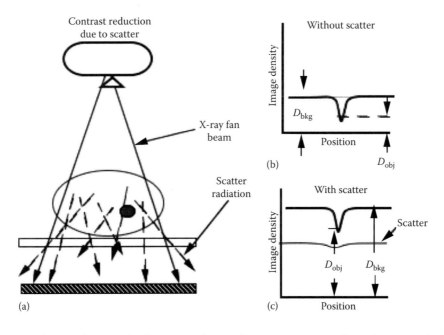

FIGURE 4.14   Scatter changes the direction of x-ray photons originating from the x-ray tube (a) producing a nearly uniform image. Contrast without scatter (b) is reduced due to such scatter (c).

Each small volume within the body is potentially a new source of x-rays, each casting a different shadow. As such, scattered x-rays strike the image receptor from various directions and carry little useful information about the object, unlike primary x-ray photons. A useful way to describe the amount of scatter in a radiographic signal is the scatter fraction F defined as

$$F = \frac{\text{Number of scattered x-ray photons}}{\text{Number of primary plus scattered photons}}. \tag{4.4}$$

Sorenson has shown that the contrast reduction due to scatter radiation is related to the scatter fraction F according to the equation

$$C_{sc} = C(1 - F), \tag{4.5}$$

where
   C_{sc} is the contrast in the presence of scatter
   C is the contrast with no scatter
   F is the scatter fraction

When the scatter fraction is 90%, as it can be in chest radiology (broad x-ray beam), the reduction in contrast due to scatter is also 90% and one obtains only 10% of the contrast that is available in the image without scatter (Figure 4.14). It is, therefore, important to reduce the amount of scatter for any imaging procedure.

There are a number of ways to reduce scattered radiation to improve radiographic contrast (Figure 4.15). We will discuss these techniques in detail in later sections and only review them conceptually here. Scatter can be reduced by geometrical means, using a small field size or by having a large "air gap" between the patient and the image receptor. In these cases, because scatter does not travel in the direction of the primary x-ray beam, it has a lower probability of striking the image receptor with a small field size or large distance (air gap) between the patient and the detector. In addition, when a small field size is irradiated, less tissue volume is exposed, and therefore, less scatter radiation is produced. It is always good radiological practice to limit the radiation exposure to as small an area (field of view) as possible, consistent with the clinical needs of the study. This reduces the radiation dose delivered to the patient and helps to recover the contrast lost due to scattered x-rays.

There are several devices that can be used to reduce scatter. The most common one is called an x-ray grid that is analogous in concept to venetian blinds (Figure 4.15b). A grid contains thin parallel strips of lead embedded in aluminum or plastic with the strips aligned with the direction of the primary x-ray beam. The grid is placed between the patient and the image receptor. This arrangement favors transmission of primary x-ray photons, while attenuating scattered photons. The grid may be moved back and forth across the lead strips during an exposure to blur out the grid pattern.

Another approach uses a scanning slit system (Figure 4.15c), which translates a pair of slits in radiopaque metal plates across the patient during the x-ray exposure. One slit

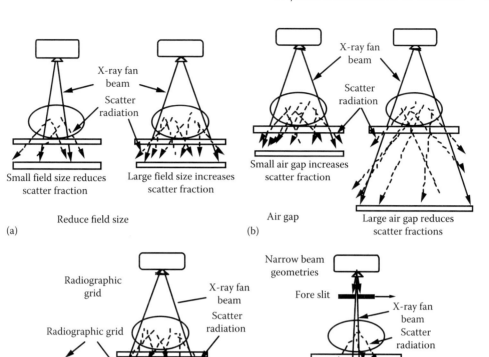

FIGURE 4.15 There are numerous approaches to reducing scattered radiation: (a) Using smallest beam width possible (upper left), (b) increasing the air gap between the patient and the detector, (c) use of radiographic grids, and (d) using a scanned narrow x-ray beam.

is before (forming a narrow beam) and the other after the patient (rejecting scatter). The narrow beam geometry provided by this arrangement of slits greatly increases the ratio of primary to scattered x-ray photons. Unfortunately, this system is not useful when short exposures are needed (to deal with motion), and it requires more output from the x-ray tube, thereby shortening tube life since only a small fraction of x-rays produced are used in forming an image. These limitations outweigh the advantages of the reduced scatter, so narrow-beam scanning systems have not proved useful for most clinical studies. However, the narrow beam geometry intrinsic to an x-ray CT system can be used to make low-scatter images if desired, but spatial resolution may not be adequate.

4.6.2 Image Intensifier Veiling Glare

Veiling glare arises from scattered x-rays at the input and output windows of an image intensifier as well as scattering of visible light in the input and output optics of the image intensifier and video camera (Figure 4.16).

Veiling glare occurs to a certain extent in many electro-optical imaging systems. Its effect is similar to that of scattered radiation in that it reduces contrast. Its contrast-reducing effect is quantified using the contrast ratio (R), which is defined as the output at the center of the image divided by the output with 10% of the II's input area blocked by

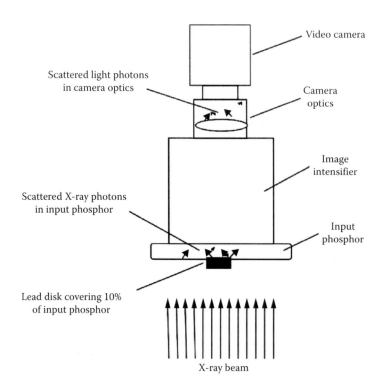

FIGURE 4.16 The setup for measuring image intensifier veiling glare using a lead disk to block x-rays at the center of the field of view.

lead (so that the output is due entirely to veiling glare—Figure 4.16). Ideally, one wants no veiling glare, corresponding to an infinite contrast ratio. Image intensifiers using cesium iodide (CsI) phosphors have a contrast ratio of 17:1, while improved systems with a fiber-optic output window and very thin titanium input window can have contrast ratios of 35:1. If an image intensifier has a contrast ratio of R, the input radiographic contrast C will be reduced to a contrast C_{vg} by veiling glare where

$$C_{vg} = C\left(\frac{R}{R+1}\right). \tag{4.6}$$

For example, if an image intensifier has a contrast ratio of 20:1, then an ideal contrast of 10% will be reduced to ~9.5% (~5% loss) due to the effects of veiling glare.

4.7 PHYSICAL DETERMINANTS OF CONTRAST FOR OTHER IMAGING MODALITIES

4.7.1 Planar Nuclear Medicine Imaging

Imaging in nuclear medicine is referred to as functional imaging. Factors that affect contrast are broadly classified as subject or detector dependent. The basic imaging device is called a gamma camera and images are called scans even when scanning is not part of the imaging protocol, such as for planar imaging. During a routine nuclear medicine imaging exam, multiple planar images are often acquired, while the camera is viewing the patient from different directions. These views are generally labeled based on the anatomical side of the body where the camera was placed, such as anterior, posterior left lateral, left anterior oblique, etc.

4.7.1.1 Subject-Dependent Contrast Factors

In nuclear medicine studies, the affinity of a target organ for a biological molecule is evaluated using a radionuclide attached to the molecule as a radioactive tracer or radiotracer (Figure 4.17). The distribution and time course of the radiotracer is recorded using a gamma camera following intravenous administration of the radiotracer. The radiotracer accumulation in the target organ relative to surrounding tissues provides the much needed contrast. This relative accumulation is called the target-to-background ratio and contrast is just the relative difference in detected activity between the target and the background. Nuclear medicine physicians are trained to recognize normal and abnormal radiotracer uptake, accumulation, and clearance patterns for various organ systems. Additionally,

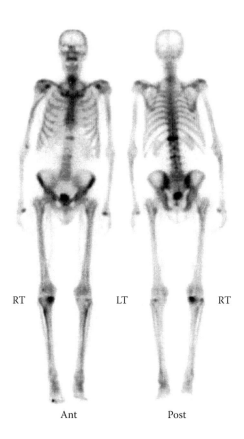

RT   LT   RT

Ant   Post

FIGURE 4.17 Whole-body bone scan of showing the distribution of the radiotracer Tc-99(m) methylenediphosphonate or MDP that localizes in bone. (Courtesy of IAEA Human Health Campus, Vienna, Austria.)

dynamic imaging (multiple images acquired over time), coupled with region of interest analysis, can be used to quantify uptake and clearance rates for comparison with normal values.

Another subject-related contrast factor is the attenuation of gammas emitted from the organ of interest by both absorption and scattering. The most common radionuclide used is Tc-99(m) with a gamma energy of 140 keV, which has a half-value layer of approximately 4.6 cm in the body. The attenuation depends on the depth of the organ of interest. For a shallow organ system such as bone, the attenuation is low compared with the liver, which extends deep into the abdomen. This attenuation effect is seen in the differences between the posterior and anterior views of the bone scan in Figure 4.17. As attenuation increases, the fraction of scatter increases, reducing subject contrast. Unfortunately, as body size increases, attenuation increases, reducing contrast.

4.7.1.2 Detector-Dependent Contrast Factors
A schematic of the main components of the most common gamma camera is provided in Figure 4.18. This gamma camera is sometimes called an Anger camera, named after its inventor Hal Anger. The gamma camera components that influence the contrast are the crystal, the collimator, and the pulse height analyzer (PHA) (spectrometer).

4.7.1.3 Crystal
The thickness of the NaI(Tl) crystal is important since a thicker crystal can absorb more gammas; however, the increased thickness leads to more uncertainty in position, and that reduces resolution and contrast for smaller objects. The energy resolution of the crystal is important to reject scatter. Energy resolution depends on the number of light photons produced on average per gamma absorbed. Energy resolution increases with increasing gamma energy. The energy resolution for Tc-99(m) is approximately 10%. The size of PMTs also affects spatial resolution with smaller PMTs providing higher resolution and therefore more contrast for smaller objects.

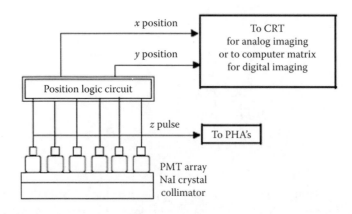

FIGURE 4.18   Diagram of Anger-type gamma camera.

4.7.1.4 Collimator

Unlike x-ray imaging where the image is produced by radiation transmitted through the body, for nuclear medicine, the image is produced by radiation that is emitted from within the body. A 2-D (planar) image of the 3-D distribution of radiotracer is formed using a device called a collimator. Collimators are made of lead with parallel channels (openings) that are perpendicular to the plane of the crystal (Figure 4.19). Gammas traveling parallel to the channels form a 2-D image of the radiotracer distribution at the crystal. These gammas must be emitted in a very small solid angle to traverse the channels, so only a small fraction of gammas emitted are used to form the image ($\sim$1 in 10^{-4}). Collimators are optimized to work within certain energy ranges, but most collimators are made for use with the 140 keV photons emitted by Tc-99(m). Collimator design is a trade-off between channel length, channel cross-sectional area, and septal thickness (Pb between channels). Resolution and sensitivity are tightly coupled with collimator designs where increasing spatial resolution diminishes sensitivity. For this reason, collimators are designed that best meet various imaging needs. When resolution is more important, a low-energy high-resolution (LEHR) collimator is preferred. Alternatively, when sensitivity is more important, a low-energy high-sensitivity (LEHS) collimator is preferred. A compromise collimator is also provided by most manufacturers, sometimes called a low-energy all-purpose (LEAP) collimator. The LEHR collimator is used when the radiotracer uptake has sharp borders, such as for extremity bone scanning. The LEHS collimator is used to improve sensitivity when tracking the entry and/or exit of a bolus of radiotracer within an organ, such as for first-pass cardiac studies. The LEAP collimator is used when there is a need for moderate spatial resolution and sensitivity.

FIGURE 4.19 Hexagonal section from a low-energy collimator used with TC-99(m). The spacing between collimator channels is ~1.9 mm for this collimator.

Resolution diminishes with increasing distance from a collimator, but sensitivity remains relatively constant. This loss of resolution with distance reduces contrast for small deep lesions. This effect is also seen in the anterior and posterior view bone scans (lumbar spine in Figure 4.17). The degradation in resolution with distance is usually less severe with an LEHR collimator compared to LEHS or LEAP collimators, so it is likely the best choice for small deep lesions. This assumes that sensitivity is not a limiting factor for the study.

4.7.1.5 Spectrometer

An energy spectrometer uses a PHA to selectively accept the energies of gammas emitted by the radiotracer. This serves to reduce the effects of background and scattered radiation. A spectrometer is generally adjusted to have a window of acceptance of ~20% of the gamma's photopeak energy. For TC-99(m), this is ±14 keV. Most spectrometers have the capability to support multiple energy windows for radiotracers such as Ga-67 and Tl-201.

Scattered radiation has a lower energy than primary radiation, so a narrow spectrometer window can be used to reject some scattered gammas. For higher-energy gammas, the change in energy is larger for the same scattering angle (basic physics of Compton scattering), so scatter reduction by a narrow window can be more effective. Also, the probability for attenuation within the body decreases with increasing energy, providing the potential for improved contrast. However, the probability of interaction within the crystal drops with increasing gamma energy, potentially leading to reduced contrast. The trade-off is to use a radionuclide with gamma energy of around 150 keV that can be collimated, that is absorbed with high efficiency with common crystal thicknesses, and that escapes the body with an acceptable level of attenuation. Tc-99(m) with its 140 keV gamma has become the radionuclide of choice.

4.7.1.6 Solid-State Gamma Cameras

Gamma cameras with designs similar to the digital imaging plates used in planar x-ray imaging with much thicker crystals are becoming popular. One system uses an array of over 11,000 tightly spaced CsI(Tl)/silicon photodiode modules in a UFOV of ~12 × 15 in., consistent with the requirements of many clinical studies. The crystal is of dimensions ~3 × 3 × 6 mm, providing intrinsic absorption efficiency similar to common 9.5 mm thick NaI(Tl) gamma cameras (~88% vs. ~90% for Tc-99(m)). The intrinsic spatial resolution is mostly determined by the crystal 3 × 3 mm dimensions and is reported to be ~3.2 mm. The energy resolution for solid-state systems with Tc-99(m) is actually slightly better than conventional gamma camera systems (6%–8% vs. 10%). Also, the count rate capability is good, maxing out at ~5 million counts per second. A one-crystal-per-pixel design where each crystal has an assigned pixel location is much simpler than the PMT position logic circuits used in older gamma cameras. This multicrystal design supports smaller, lighter, and potentially more mechanically robust gamma cameras.

Other solid-state designs are available, some dedicated to mammography and some even proposed for handheld imaging. Collimators and spectrometers for solid-state gamma cameras are similar to those used for conventional gamma cameras, so operators readily adapt to their use.

4.7.2 Magnetic Resonance Imaging

Most magnetic resonance images (MRIs) are based on signals from hydrogen protons in tissue, more specifically from mobile water molecules. The magnetic moment of the hydrogen protons and the magnet field strength makes these protons sensitive to RF at a frequency of ~43 MHz/T, the resonance frequency (called the Larmor frequency). An RF pulse at the Larmor frequency is used to alter the net magnetization of the protons so that they produce a measurable RF signal. The potential signal strength depends on relaxation times (T1 and T2) and proton density (PD) of tissues. MR imaging requires exquisite timing between RF transmitter and receiver as well as the applied magnetic field gradients. The simultaneous timing of this is programmed into and executed by the MR pulse sequencer. A common clinical MRI uses a pulse sequence classified as spin-echo (SE) (Figures 4.20 and 4.21). Like nuclear medicine imaging, we can separate factors that affect contrast for SE imaging into subject- and detector-dependent factors.

4.7.2.1 Subject-Dependent Contrast Factors

The MRI signal is proportional to PD, the concentration of mobile water molecules. PD is lowest for air, low for bone, high for water, and similar to that of water (but lower) for most tissues. There are two relaxation processes that affect MRI signals, with spin–lattice relaxation quantified by relaxation time T1 and spin–spin relaxation by T2. The T1 relaxation time varies with magnetic field strength (B_0) but more importantly by tissue, which produces T1W contrast. In the brain, T1 is longest for CSF, shortest for fat, and intermediate for other tissues. Unlike T1, the T2 relaxation time is minimally affected by field strength. T2 is much shorter than T1 (approximately 1/10th). The T2 relaxation time varies by tissue with T2 relaxation times increasing as T1 relaxation

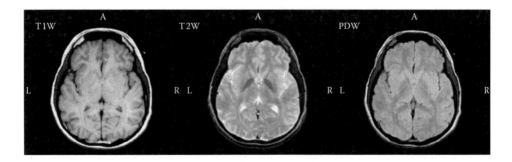

FIGURE 4.20 MR images acquired to produce image-specific contrast between soft tissues. The T1-weighted (T1W) image contrast is mostly due to differences in T1, while in the T2- and PD-weighted images (T2W and PDW), the contrast is mostly due to differences in T2 or PD. All three images were acquired using SE-type pulse sequences.

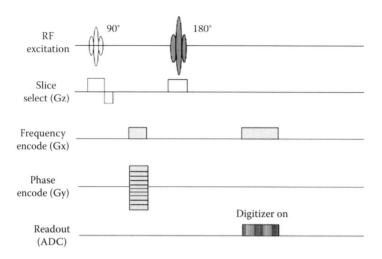

FIGURE 4.21   SE pulse sequence diagram. The lines of this diagram are greatly simplified but show the timing of the systems RF, gradient, and digitizer. The sequence repeats many times with a period called the repetition time (TR). TE is time from 90° RF pulse to middle of the RF echo seen in the top line. Note that the signal that is received (digitized) coincides with the RF echo.

times increased. Potential contrast between tissues depends on the tissue differences in T1, T2, and PD. There are large differences in relaxation times for tissues in the brain (gray matter, white matter, and CSF) that can provide high image contrast between these tissues (Figure 4.20).

4.7.2.2 Detector-Dependent Contrast Factors

Similar to nuclear medicine imaging, the signals from MR images are intrinsic, arising from within the body. The signals produced are called RF signals since they fall within the radio-frequency band of the radiation spectrum. Unlike nuclear medicine, there is no ionizing radiation and little attenuation associated with the RF signals. The imaging system consists of an RF transmitter, RF coils, gradient coils for spatial encoding, and an RF receiver. The signals from the RF receiver are digitized and sorted into "k-space," which is a Fourier transform of signals from the object (subject or patient). The magnitude portion of the inverse Fourier transform of the k-space image is used to form MR images.

For SE MR imaging, contrast weighting is controlled by the operator-selected acquisition parameter's repetition time (TR), echo time (TE), pixel spacing, and slice thickness. Pixel spacing and slice thickness determine the size of volume elements (voxels) in the images. The voxel volume is roughly the volume of space from which signals arise, so smaller voxels are needed for higher spatial resolution. A typical SE pulse sequence diagram shows the relationships among time of the RF excitation, the three gradients used, and the analog-to-digital converter (ADC) (Figure 4.21).

While T1 or T2 relaxation times can be calculated for different tissues, they require multiple images while varying TR and TE and fitting to relaxation models for each voxel.

TABLE 4.5   Weighted Contrast in Spin-Echo MR Images

Acquisition Parameter	T1W	T2W	PDW
TR	Short	Long	Long
(ms)	(450–850)	(2000+)	(2000+)
TE	Short	Long	Short
(ms)	(10–30)	(>60)	(10–30)

However, it is possible to acquire individual images with contrast weighted by either T1 (T1W) or T2 (T2W) relaxation times, or proton density (PDW), without model fitting. Table 4.5 summarizes how TR and TE can be set up to develop weighted contrast images such as those in Figure 4.20 for a 2–3 T MR imager. Note that short TR in combination with long TE is not used in SE imaging.

In T1W images, tissues with shorter T1 times (e.g., fat) have higher signals. In T2W images, tissues with longer T2 times (e.g., CSF) have higher signals. Edema associated with trauma also has a higher signal in T2W images. In PDW images, tissues with greater proton densities have higher signals.

4.7.3 X-Ray Computed Tomography

Unlike a planar x-ray image that is a map of the x-ray's intensity transmitted through a body, an x-ray CT image is derived from a map of linear attenuation coefficients for each volume element (voxel) in the reconstructed image. Importantly, contrast between soft tissues is greatly improved in x-ray CT due to the removal of overlying and underlying tissues.

4.7.3.1 Subject-Dependent Contrast Factors

Subject contrast for x-ray CT imaging is due to differences in linear attenuation coefficients. The subject contrast between neighboring tissues with linear attenuation coefficients μ_1 and μ_2 is calculated as

$$\text{Contrast} = (\mu_1/\mu_2 - 1) = [(\mu_m \rho)_1/(\mu_m \rho)_2 - 1]. \tag{4.7}$$

As for planar radiography, contrast between bone tissue and surrounding tissues is high. Inspection of Equation 4.7 reveals that there are two special cases where subject contrast is dependent on either difference in densities or mass attenuation coefficients.

Case 1: If $\mu_{m1} \approx \mu_{m2}$, then contrast $\approx (\rho_1/\rho_2 - 1)$ only due to density differences.

Case 2: If $\rho_1 \approx \rho_2$, then contrast $\approx (\mu_{m1}/\mu_{m2} - 1)$ only due to mass attenuation differences.

Since attenuation coefficients vary with energy, a process that affects subject contrast is beam hardening, increasing of the average energy of the x-ray beam deeper within the subject. This potentially leads to differences in calculated linear attenuation coefficients for the same tissues in thin versus thick body regions. To help compensate for this, CT images are acquired using a high 125 kVp with additional Al filtering to effectively preharden the beam.

4.7.3.2 Detector-Dependent Contrast Factors

Most CT scanners are operated in a multislice mode where data from multiple slices are collected simultaneously. An array of detectors is used, with many (up to 64) for the between-slice direction and more for the within-slice direction (800 or more). A collimator is used on the x-ray tube side of a CT scanner to limit the x-ray beam to the extent of the detectors during scans. Additionally, scanners follow a helical or spiral scan mode where the x-ray tube and detectors are mounted on a rotating gantry using slip-ring technology and the patient bed is simultaneously moving during the scan. Unlike single-slice detectors with narrow x-ray beams, the large volume of the x-ray beam used with multislice detectors leads to more scatter. Multifinned septa are used to reduce acceptance of this scattered radiation. Detector size in the between-slice direction determines the slice thickness. Slice thickness and pixel size determine voxel size, and smaller voxels are needed to preserve contrast of small objects and at borders between bone and soft tissues. Operators can select slice thickness and reconstructed pixel sizes to adapt to the needs of the CT exam, but slice thickness is the primary concern.

4.7.3.3 Special Considerations for X-Ray CT Contrast

To standardize values for x-ray CT images, the linear attenuation coefficients are converted to CT#'s:

$$CT\# = 1000 \, (\mu - \mu_w)/\mu_w. \tag{4.8}$$

Inspection of this equation shows that the CT# is a measure of the contrast between a tissue's linear attenuation coefficient and that of water. The use of CT#'s helps to reduce variability in reported values between x-ray CTs from different manufacturers and different generations of CT systems. CT#s range from −1000 for air, slightly negative for fat, zero for water, slightly positive for soft tissues, to greater than 2000 for dense bone (Figure 4.22).

We cannot calculate image contrast between CT#s since they are not on a zero-based scale. We can shift CT#s to run from zero up by adding 1000 and then calculate contrast. If we do this, we get the same result as in Equation 4.7, that in x-ray CT the intrinsic contrast is the subject contrast due to differences in linear attenuation coefficients. Also, unlike planar x-ray imaging, the thickness of the target tissue is not an explicit variable in CT contrast. This holds as long as the target tissue is larger than the voxels. Subject contrast can be modified for x-ray CT using contrast agents similar to what is done for planar imaging. However, these are usually administered intravenously or

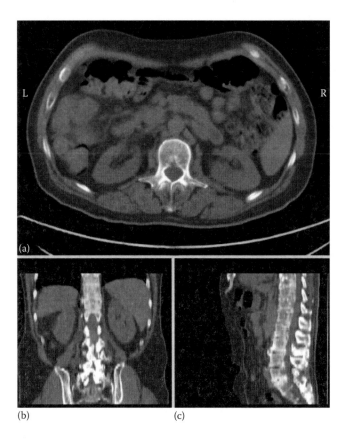

FIGURE 4.22 X-ray CT of abdomen with axial (a), coronal (b), and sagittal (c) sections illustrated. The CT# display range was set to −215 to 457 to cover tissues of interest.

orally with a delay before imaging to allow for distribution throughout the vasculature and various tissues.

In clinical practice, the contrast in CT images is determined by display settings, referred to as window and level. For example, a window can be set to provide contrast across fat and soft tissues (Figure 4.22). A narrow window provides more contrast and a broader window provides more latitude. The window level determines which CT# is at the midgray level and the window width determines the range of CT#s that fall between displayed black and white levels.

If a voxel contains multiple tissues, this impacts contrast through a physical process called the partial volume averaging. The resulting CT# will fall between that of the tissue with the lowest value and that with the highest. Larger voxels have greater potential for partial volume averaging and affect contrast at boundaries between tissues. Another source of volume averaging is associated with low-pass filtering that may be used to reduce noise levels. As the spatial extent of the low-pass filter is increased, partial volume averaging increases. Such filters are chosen as a compromise between reducing noise levels and reducing spatial resolution.

4A APPENDIX

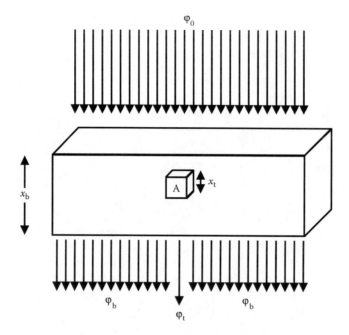

The figure on the right is the basis for developing a simple mathematical model for calculation of contrast. Model limitations include failure to account for spatial resolution and scatter, but these can be added later. This simple model will be used throughout the course, so it is important to understand how it was derived and how it can be modified to account for other factors that are important to medical imaging.

The mathematical model given in Equation 4.2 for radiographic contrast based on physical characteristics of tissues (C_p) is

$$C_p = e^{-(\mu_t - \mu_b)x_t} - 1,$$
(4A.1)

where
μ_t is the linear attenuation coefficient of target tissue
μ_b is the linear attenuation coefficient of background tissue
x_t is the thickness of target tissue of interest along x-ray beam

Given the three physical parameters for this equation, we can calculate the radiographic contrast (C_p) between two tissues, that is, prior to the detector. If μ_t is equal to μ_b, then contrast $C_p = 0$, as expected. If μ_t is larger than μ_b, then contrast C_p is negative but cannot exceed –1. Conversely, if μ_t is smaller than μ_b, then contrast C_p is positive and can exceed unity. This can occur if air is the target tissue or bone is the background tissue, but we will be more concerned with cases where these two linear attenuation coefficients are nearly equal (comparing soft tissue to soft tissue or vessels with contrast material).

In Chapter 1, we introduced the Rose model equation to provide a relationship between contrast at the threshold of visual detection (C_d) and several other physical parameters of interest:

$$k^2 = C_d^2 \varphi_b A_t, \qquad (4A.2)$$

where
  k is the theoretical value of SNR at detection threshold ($k \sim 5$–7)
  φ_b is the photon fluence (photons/area) in background
  A_t is the cross-sectional area of target tissue

Note that C_d is squared so that positive and negative values of contrast are treated the same.

By setting C_d (from Equation 4A.2) = C_p (from Equation 4A.1), we get an expression for the relationship between five physical variables at the visual threshold of detection for a target tissue.

$$k^2 = \left[e^{-(\mu_t - \mu_b)x_t} - 1 \right]^2 \varphi_b A_t. \qquad (4A.3)$$

This equation can be solved for any single physical variable, if other variables are known.

It is often of scientific and/or clinical interest to estimate the incident radiation exposure to visualize a small lesion. To do this, you will need to know the incident photon fluence φ_0 and the energy (E) of the incident photons. For x-ray beams, we usually substitute the mean energy of the beam that ranges from 40 to 50 keV for most studies.

The following equation can be used to calculate radiation exposure in mR:

$$X[\mathrm{mR}] = 1.833 \times 10^{-8} \, \frac{\mathrm{g\text{-}mR}}{\mathrm{keV}} \times \varphi_0 \left[\frac{\mathrm{photons}}{\mathrm{cm}^2} \right] \times E \left[\frac{\mathrm{keV}}{\mathrm{photon}} \right] \times \left(\frac{\mu_{en}}{\rho} \right)_{air} \left[\frac{\mathrm{cm}^2}{\mathrm{g}} \right]. \qquad (4A.4)$$

Values for the mass energy absorption coefficient of air can be found as a function of energy in various radiological physics texts as well as the handbook of radiological health. Table 4A.1 can be used for many interesting problems that deal with dose calculations.

TABLE 4A.1   Mass and Linear Attenuation Coefficients

Photon Energy (keV)	$(\mu_{en}/\rho)_{air}$ (cm²/g)	μ_{water} (cm⁻¹)
40	0.0625	0.2629
50	0.0382	0.2245
60	0.0289	0.2046

TABLE 4A.2   Physical Characteristics of Common Elements in the Body

Element	Z	A	Z/A	Density (g/cm³)	K-Edge (keV)
H	1	1	1.00	0.0586	0.01
C	6	12	0.50	2.25	0.24
O	8	16	0.50	1.14	0.53
N	7	14	0.50	0.808	0.39
P	15	31	0.48	1.82	2.14
Ca	20	40	0.50	1.55	4.04

Example using Equation 4A.4. A chest film might require an incident exposure of 20 mR using an x-ray beam with an effective energy of 50 keV. What is the incident photon fluence?

$$\varphi_0 = \frac{20 \text{ mR}}{1.833 \times 10^{-8} \dfrac{\text{g-mR}}{\text{keV}} \times 50 \dfrac{\text{keV}}{\text{photon}} \times 0.0382 \dfrac{\text{cm}^2}{\text{g}}} = 5.71 \times 10^8 \dfrac{\text{photons}}{\text{cm}^2}$$

The following material should be helpful when reviewing the material in Table 4A.2:

The electron density [electrons/g] $= N_0 \dfrac{Z}{A}$,

where

N_0 is Avogadro's number ~6.023×10^{23} (atoms/gram-atom)
Z is the atomic number (number of electrons per atom)
A is the atomic mass (grams/gram-atom)

Note that electron density is proportional to Z/A.

The effective atomic number (Z_{eff}) is estimated using an equation like the following:

$$Z_{eff} = \sqrt[2.94]{\alpha_1 Z_1^{2.94} + \alpha_2 Z_2^{2.94} + \alpha_3 Z_3^{2.94} + \cdots}, \tag{4A.5}$$

where

α_n is the fraction of electrons from element 1...n
Z_n is the atomic number of element 1...n

HOMEWORK PROBLEMS

P4.1 Assume that a 1 mm diameter artery contains a 10 mg/cm³-concentration iodine solution.

(a) Calculate the approximate radiographic (subject) contrast assuming an effective attenuation coefficient of 15 cm²/g if no scatter is present. Make any reasonable assumptions you need to answer this question.

(b) What is the radiographic contrast if the scatter fraction is 0.5?

P4.2 In a quantum-limited imaging system, a low-contrast object is just distinguishable from its surroundings.

(a) If the object has an area of 1 cm² and the measured exposure is 10^{-5} R, use the Rose model to estimate the radiographic contrast. Make any reasonable assumptions you need to answer this question.

(b) If the object has an attenuation coefficient of 10 cm²/g, estimate the product of its density and thickness that is just distinguishable from its surroundings.

P4.3 Consider two imaging situations using contrast agents with electronic detectors. In one case, a detector of cesium is used to image an iodine contrast agent. In the other case, a detector of iodine is used to image a cesium contrast agent.

(a) In which case would you expect the better image contrast? Why?

(b) In which case would you expect the better signal-to-noise ratio at a given patient exposure level? Why?

(c) Which of these two hypothetical situations is more realistic in terms of commonly available contrast agents and detector materials?

P4.4 Polycythemia is a disease characterized by an abnormal increase in the number of circulating red blood cells. In its normal function of breaking down hemoglobin from dead red blood cells, the livers of patients with polycythemia accumulate abnormal amounts of iron.

(a) Ignoring patient radiation dose considerations and knowing that the K-edge energy of iron is 7.11 keV, describe how you would choose the kVp and the filter for the x-ray tube to maximize liver contrast for patients with polycythemia.

(b) Briefly comment whether your answer in (a) is clinically realistic when patient radiation dose is considered. A short calculation may help to clarify your answer.

P4.5 Digital subtraction angiography (DSA) is a technique using a digital fluorographic system. In DSA, one image (the "mask image") is obtained before an iodinated contrast agent is injected into the circulatory system. A second image (the "opacification image") then is obtained after injection of the contrast agent. The images are logarithmically transformed, then subtracted to remove anatomical background structures to isolate the contrast agent in the circulatory system (typically arteries and cardiac chambers). The geometry of the two images is summarized in the following diagrams.

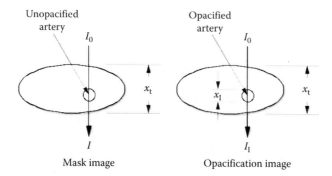

where
I_0 is the incident photon fluence for both the mask and opacification images
I is the photon fluence detected in the mask image
I_I is the photon fluence detected in the opacification image
μ_I is the linear attenuation coefficient of iodine within the artery
x_I is the equivalent thickness of iodine from the contrast agent in the opacification image
μ_t is the linear attenuation coefficient of tissue
x_t is the equivalent thickness of tissue in both the mask and opacification images

(a) Show that for the mask image

$$I = I_0 e^{-\mu_t x_t}.$$

and that for the opacification image

$$I_I = I_0 e^{-(\mu_t x_t + \mu_I x_I)}.$$

(b) If we define the logarithmic difference image L to be

$$L = \ln(I_I) - \ln(I),$$

show that

$$L = -\mu_I x_I.$$

Briefly discuss assumptions and how this result shows that logarithmic subtraction removes anatomical background structures from the DSA image.

(c) Define the scatter-to-primary ratio f as

$$f = \frac{\text{Exposure contributed by scattered radiation}}{\text{Exposure contributed by primary radiation}}.$$

In the presence of scatter radiation, if f is the scatter-to-primary ratio, the photon fluence in the mask image is contributed by both the primary (unscattered) and the scattered components. In this case, show that the detected photon fluence is

$$I = I_0 e^{-\mu_t x_t} + f \cdot I_0 e^{-\mu_t x_t}.$$

Assuming that the scatter field does not change significantly by the addition of a small amount of iodine, show that for the opacification image,

$$I = I_0 e^{-(\mu_t x_t + \mu_I x_I)} + f \cdot I_0 e^{-\mu_t x_t}.$$

(*Hint*: Part (c) follows directly from the definition of the scatter-to-primary ratio. The solution is very simple. Do not make it too difficult for yourself.)

(d) For a scatter-to-primary ratio f, show that in the presence of scatter, the logarithmic difference image is

$$L = \ln\left(\frac{e^{-u_I x_I} + f}{1 + f}\right)$$

which reduces to

$$L \cong -\frac{\mu_I x_I}{1 + f}$$

for small values of $\mu_I x_I$. Discuss the importance of this result in terms of the image contrast that is obtained in DSA.

P4.6 Dual-energy imaging is a technique in which two images are obtained with different energy spectra with the purpose of separating two materials having dissimilar atomic numbers that are both present in the imaging field. In most realistic systems, the spectra used for image acquisition are polyenergetic bremsstrahlung spectra, but for purposes of this problem, we can approximate the beams as being monoenergetic with energies E_1 and E_2. Assume that

I_{01} is the incident photon fluence at energy E_1
I_1 is the detected photon fluence at energy E_1
I_{02} is the incident photon fluence at energy E_2
I_2 is the detected photon fluence at energy E_2

(a) If the imaging field contains only bone (b) and soft tissue (t), and if

μ_{b1} is the linear attenuation coefficient of bone at energy 1
μ_{b2} is the linear attenuation coefficient of bone at energy 2
μ_{t1} is the linear attenuation coefficient of soft tissue at energy 1
μ_{t2} is the linear attenuation coefficient of soft tissue at energy 2

show that the thickness of bone x_b and the thickness of soft tissue x_t are equal to

$$x_b = \frac{\mu_{t1}\ln\left(\dfrac{I_{02}}{I_2}\right) - \mu_{t2}\ln\left(\dfrac{I_{01}}{I_1}\right)}{\mu_{t1}\mu_{b2} - \mu_{t2}\mu_{b1}},$$

$$x_t = \frac{\mu_{b1}\ln\left(\dfrac{I_{02}}{I_2}\right) - \mu_{b2}\ln\left(\dfrac{I_{01}}{I_1}\right)}{\mu_{b1}\mu_{t2} - \mu_{b2}\mu_{t1}}.$$

(b) Briefly discuss how the results in (a) allow you to separate bone and soft tissue when imaging the human body using dual-energy techniques.

CHAPTER 5

Mathematics for Linear Systems

I N THIS CHAPTER, WE review the concepts of linear transformations and position invariant functions as they relate to imaging. With some constraints, an image can be represented as the convolution of a position-invariant point spread function within the object, supporting utilization of the Fourier transform theory to alternatively describe imaging. An important factor for digital imaging is sampling, so we will examine the mathematics of sampling theory and model the effects of digitization of the medical image. Mathematical template functions and associated tools presented in this chapter will be used throughout the remainder of the text, and they should be familiar to the graduate-level engineer or physicist, so only a brief overview will be presented. Those who need review or further explanation of these topics are referred to several excellent texts.*

5.1 DIRAC DELTA FUNCTION

We start with the Dirac delta function $\delta(x)$, which is defined as follows:

$$\delta(x) = \begin{cases} +\infty & \text{if } x = 0 \\ 0 & \text{if } x \neq 0 \end{cases}. \tag{5.1}$$

Importantly, this has a unity integral value that is naturally unitless:

$$\int\limits_{-\infty}^{+\infty} \delta(x)\,dx = 1. \tag{5.2}$$

---

* An excellent discussion is presented in Ronald N. Bracewell, *The Fourier Transform and Its Applications*, New York: McGraw-Hill Book Company, 2nd edn., c. 1986. A shorter but equally commendable discussion focusing on medical imaging is presented in Albert Macovski, *Medical Imaging*, Englewood Cliffs, NJ: Prentice-Hall, c. 1983. A thorough discussion of linear systems in medical imaging is given in the two-volume text Harrison H. Barrett and William Swindell, *Radiological Imaging: The Theory of Image Formation, Detection, and Processing*, New York: Academic Press, c. 1981.

We will call $\delta(x)$ a "delta function," with implicit rather than explicit homage to Dirac. This delta function has several important interesting properties. First, it is rather easy to show that the delta function greatly simplifies integration when it is present in the integrand with another function:

$$f(0) = \int_{-\infty}^{+\infty} f(x)\delta(x)\mathrm{d}x \tag{5.3}$$

This equation states that when the product of a function and the delta function is integrated over the domain of the function, the result is the value of the function at $x = 0$, assuming it exists. This property can be generalized to provide a sample of $f(x)$ at position $x = x'$ as follows:

$$f(x') = \int_{-\infty}^{+\infty} f(x)\delta(x - x')\mathrm{d}x \tag{5.4}$$

where $\delta(x - x')$ is a shifted delta function, shifted from $x = 0$ to $x = x'$.

A series of delta functions is called a "comb" or "shah" function, and these can be represented as a discrete sum since they do not overlap:

$$\mathrm{III}(x) = \sum_{n=-\infty}^{+\infty} \delta(x - n) \tag{5.5}$$

Equation 5.5 is an infinite sum of delta functions, which are regularly spaced at integer intervals (*Note*: n and x have the same units). A function can be "point sampled" at integer intervals if it is multiplied by the comb function:

$$f(x)\mathrm{III}(x) = \sum_{n=-\infty}^{+\infty} f(x)\delta(x - n) \tag{5.6}$$

so that the integral of this product yields the sum of nonoverlapping sampled function values $f(x)$ where $x = n$:

$$\int_{-\infty}^{+\infty} f(x)\mathrm{III}(x)\mathrm{d}x = \sum_{n=-\infty}^{\infty} f(n) \tag{5.7}$$

We will return to sampling and the delta function after we review a few more mathematical tools.

5.2 CONVOLUTION

The convolution of $f(x)$ with $h(x)$ can be written using the convolution symbol "$\otimes$" as $f(x) \otimes h(x)$. The convolution operation is defined in the following equivalent integral equations:

$$f(x) \otimes h(x) = \int_{-\infty}^{+\infty} f(x-y)h(y)\mathrm{d}y = \int_{-\infty}^{+\infty} f(y)h(x-y)\mathrm{d}y \qquad (5.8)$$

In Equation 5.8, y is a dummy variable for x. In two dimensions, convolution is defined as follows:

$$f(x,y) \otimes \otimes h(x,y) = \int_{-\infty}^{+\infty}\int_{-\infty}^{+\infty} f(x-s,y-t)h(s,t)\mathrm{d}s\,\mathrm{d}t = \int_{-\infty}^{+\infty}\int_{-\infty}^{+\infty} f(s,t)h(x-s,y-t)\mathrm{d}s\,\mathrm{d}t \quad (5.9)$$

The symbol "$\otimes\otimes$" indicates a convolution operation on both independent variables. This seemingly awkward notation reduces the ambiguity of the expression $f(x, y) \otimes h(x, y)$, which could mean alternatively a one-dimensional convolution in terms of x, a one-dimensional convolution in terms of y, or a two-dimensional convolution in terms of both x and y. In this text, the expression $f(x, y) \otimes h(x, y)$ will denote a one-dimensional convolution, and to avoid ambiguity, we will state explicitly (or make it clear implicitly) with respect to which independent variable the convolution is to be obtained.

We can show (actually you should show) that the convolution operation has both commutative and associative properties so that the order and grouping of functions is completely interchangeable:

$$\text{Commutative property:} \quad f(x) \otimes g(x) = g(x) \otimes f(x) \qquad (5.10)$$

$$\text{Associative property:} \quad f(x) \otimes [g(x) \otimes h(x)] = [f(x) \otimes g(x)] \otimes h(x) \qquad (5.11)$$

Other interesting features of the convolution operation are its "shifting" and "replication" properties. It follows directly from the definition of a delta function at $x = a$. The *shifting* property for convolution states that

$$f(x) \otimes \delta(x - a) = f(x - a) \qquad (5.12)$$

such that a function $f(x)$ can be relocated (or shifted) by convolving it with a shifted delta-function. A function can be *shifted* by a and scaled by b using this property as follows:

$$f(x) \otimes \delta\left(\frac{x-a}{b}\right) = bf\left(x-a\right) \qquad (5.13)$$

Finally, $f(x)$ can be *replicated* an infinite number of times at unit spacing by convolving the function with the comb function:

$$f(x) \otimes \mathrm{III}(x) = \sum_{n=-\infty}^{+\infty} f(x-n) \qquad (5.14)$$

These expressions can be generalized for two-dimensional convolutions, and their proof will be left as exercises for the reader.

5.3 FOURIER TRANSFORM

In one-dimension, the Fourier transform of $f(x)$ is defined by the integral equation

$$F(u) = \int_{-\infty}^{+\infty} f(x)e^{-2\pi iux}\,dx \tag{5.15}$$

where $F(u)$ is expressed in terms of the spatial frequency variable u. Similarly, the inverse Fourier transform of $F(u)$ gives us the function $f(x)$ where

$$f(x) = \int_{-\infty}^{+\infty} F(u)e^{2\pi iux}\,du \tag{5.16}$$

Note the use of capital letters for Fourier domain functions and lower case for spatial domain functions, while keeping the function letter unchanged. The spatial frequency variable (u) has units that are the inverse of those of the spatial variable (x). So if x is in mm, then u is in 1/mm. Together $f(x)$ and $F(u)$ represent Fourier transform pairs.

In two dimensions, $f(x, y)$ and its Fourier transform $F(u, v)$ are related by the following integral equations:

$$F(u,v) = \int_{-\infty}^{+\infty}\int_{-\infty}^{+\infty} f(x,y)e^{-2\pi i(ux+vy)}\,dxdy \tag{5.17}$$

and

$$f(x,y) = \int_{-\infty}^{+\infty}\int_{-\infty}^{+\infty} F(u,v)e^{2\pi i(ux+vy)}\,dudv \tag{5.18}$$

The Fourier transform has enough interesting properties to fill books and keep the precious time of mathematicians, engineers, and physicists fully occupied. Here, we present several basic definitions and properties, and as other properties are needed to describe imaging systems, they will be introduced.

To simplify the notation, we sometimes use the symbol $\Im$ to indicate Fourier transform. We will focus on a few Fourier transforms that will be useful for imaging scientists. The first is

$$\Im\{f[x/a]\} = aF(au) \tag{5.19}$$

and if $a > 1$, we see that "expansion" in the spatial domain (division of x by a) causes "contraction" in the spatial frequency domain (multiplication of u by a). Likewise, expansion in the frequency domain results in contraction in the spatial domain. One might expect this from the fact that units in the spatial domain are "mm," while those in the frequency domain are "1/mm." This also follows in two dimensions for $a > 1$ and $b > 1$:

$$\Im\{f(x/a, y/b)\} = ab\, F(au, bv) \tag{5.20}$$

One of the most useful properties of the Fourier transform is its relationship to convolution where convolution between functions in one domain becomes multiplication of functions in the other:

$$\Im\{f(x) \otimes g(x)\} = F(u)\, G(u) \tag{5.21}$$

such that

$$f(x) \otimes g(x) = \Im^{-1}\{F(u)G(u)\} \tag{5.22}$$

and

$$\Im[f(x)\, g(x)] = F(u) \otimes G(u) \tag{5.23}$$

such that

$$f(x)\, g(x) = \Im^{-1}\{F(u) \otimes G(u)\} \tag{5.24}$$

The spatial domain convolution operation can be rather difficult to apply analytically, but is one that arises often in medical imaging. By utilizing the Fourier transform, we can perform the convolution operation by multiplying the Fourier transform of the constituent functions, then taking the inverse transform (Equation 5.22). Multiplication in the frequency domain is done as the product of two complex functions, but this operation is straightforward in most cases. *Because this approach is so utilitarian, we can perform much of the mathematics of medical imaging by learning Fourier transforms of a few basic functions used to describe imaging systems.*

Following are definitions of five common template functions that are key to our mathematical description of medical imaging:

$$\text{Delta function: } \delta(x) = \begin{cases} +\infty & \text{if } x = 0 \\ 0 & \text{if } x \neq 0 \end{cases} \quad \text{where } \int_{-\infty}^{+\infty} \delta(x)\mathrm{d}x = 1 \tag{5.25}$$

$$\text{Sinc function: } \mathrm{sinc}(x) = \frac{\sin(\pi x)}{\pi x} \tag{5.26}$$

$$\text{Rect or box function: } \mathrm{II}(x) = \begin{cases} 1 & \text{if } |x| \leq \dfrac{1}{2} \\ 0 & \text{if } |x| > \dfrac{1}{2} \end{cases} \tag{5.27}$$

$$\text{Triangle function: } \wedge(x) = \begin{cases} 1 - |x| & \text{if } |x| \leq 1 \\ 0 & \text{if } |x| > 1 \end{cases} \tag{5.28}$$

$$\text{Comb or Shah function: } III(x) = \sum_{n=-\infty}^{+\infty} \delta(x-n) \qquad (5.29)$$

Note: The delta, sinc, rect, and triangle functions are each defined such that their integral is unity.

The Fourier transform of basic template functions along with several trigonometric functions and the classic Gaussian function is of major importance in medical imaging (Table 5.1 and Figure 5.1).

TABLE 5.1 Fourier Transform Pairs Commonly Used in Medical Imaging

Spatial Domain $f(x)$	Frequency Domain $F(u)$	
$\delta\left(\dfrac{x}{a}\right)$	a	(5.30)
$\dfrac{1}{2}\left[\delta\left(x+\dfrac{a}{2}\right)+\delta\left(x-\dfrac{a}{2}\right)\right]$	$\cos(\pi au)$	(5.31)
$\dfrac{i}{2}\left[\delta\left(x+\dfrac{a}{2}\right)-\delta\left(x-\dfrac{a}{2}\right)\right]$	$-\sin(\pi au)$	(5.32)
$\cos\left(\dfrac{\pi x}{a}\right)$	$\dfrac{1}{2}\left[\delta\left(u+\dfrac{1}{2a}\right)+\delta\left(u-\dfrac{1}{2a}\right)\right]$	(5.33)
$\sin\left(\dfrac{\pi x}{a}\right)$	$\dfrac{i}{2}\left[\delta\left(u+\dfrac{1}{2a}\right)-\delta\left(u-\dfrac{1}{2a}\right)\right]$	(5.34)
$e^{-\frac{\pi x^2}{a^2}}$	$ae^{-\pi a^2 u^2}$	(5.35)
$II\left(\dfrac{x}{a}\right)$	$a\,\text{sinc}(au)$	(5.36)
$\wedge\left(\dfrac{x}{a}\right)$	$a\,\text{sinc}^2(au)$	(5.37)
$III\left(\dfrac{x}{a}\right)$	$a\,III(au)$	(5.38)
$\text{sinc}\left(\dfrac{x}{a}\right)$	$a\,II(au)$	(5.39)
$\text{sinc}^2\left(\dfrac{x}{a}\right)$	$a\wedge(au)$	(5.40)
$f\left(\dfrac{x}{a}\right)$	$aF(au)$	(5.41)
$\delta(x) = \displaystyle\int_{-\infty}^{+\infty} e^{2\pi iux}\,du$	$\delta(u) = \displaystyle\int_{-\infty}^{+\infty} e^{-2\pi iux}\,dx$	(5.42)

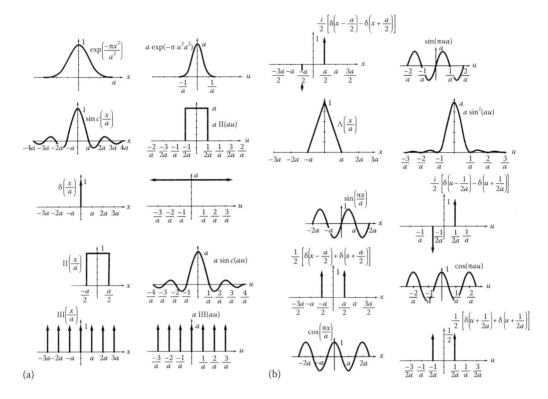

FIGURE 5.1 Graphical representation of common template functions (a) and their Fourier transforms (b).

5.4 MODELING MEDICAL IMAGING AS A LINEAR SYSTEM

The imaging process can be modeled as a system that mathematically transforms an object to a corresponding image. In a simple 2-D model, the mathematical transform (S), a function of the imaging system, determines how the object $f(x, y)$ is transformed to the image $g(x, y)$. For x-ray imaging, the object function $f(x, y)$ alters the x-ray transmission at position (x, y) in the object for a thin slab of thickness dz. We will only deal with this slab for now. The image function $g(x, y)$ is modeled as the x-ray photon fluence at (x, y) in the image (i.e., the radiographic image). The most basic form for an imaging equation is the following:

$$g(x, y) = S[f(x, y)] \tag{5.43}$$

For much of the discussion of medical imaging systems, we will assume that they are linear systems. This assumption is made implicitly, or we can force it to happen explicitly, and although it limits us to a fairly specific set of characteristics, it provides several attractive properties.

A linear imaging system defined in terms of a linear imaging system transform S has two important properties. First, if we image the sum of two objects f_1 and f_2, the result is identical to that obtained if we add the images of the objects acquired separately:

$$S[f_1(x, y) + f_2(x, y)] = S[f_1(x, y)] + S[f_2(x, y)] \tag{5.44}$$

This is known as the property of linear additivity. This allows us to model imaging for a thin slab under the assumption that if all slabs were modeled, we could sum the parts that compose the image. Second, if we image the object $f(x, y)$ and then multiply the image by a constant a, we get the same result as imaging an object of magnitude a times $f(x, y)$. This can be represented mathematically as

$$aS[f(x, y)] = S[af(x, y)] \tag{5.45}$$

where a is the multiplicative constant. This is known as the property of linear scalar. The properties of linear additivity and scaling can be summarized in a single definition of the linear operator. The operator S is said to be linear if for every two functions f_1 and f_2 in its domain and for every two constants a and b, S behaves as follows:

$$aS[f_1(x, y)] + bS[f_2(x, y)] = S[af_1(x, y) + bf_2(x, y)] \tag{5.46}$$

An example of a linear operator is the derivative operator (d/dx). That is, if f_1 and f_2 are two functions, then

$$a\frac{d}{dx}\big[f_1(x)\big] + b\frac{d}{dx}\big[f_2(x)\big] = \frac{d}{dx}\big[af_1(x) + bf_2(x)\big] \tag{5.47}$$

Other examples of linear operators are the integral as well as most smoothing and edge-sharpening operations. *Importantly, convolution and the Fourier transform are linear operations.*

Investigation of how the imaging system operator S transforms a delta function provides insight about image formation. We begin the derivation using the two-dimensional shifting property of the delta function applied to $f(x, y)$:

$$f(x,y) = \int_{-\infty}^{+\infty}\int_{-\infty}^{+\infty} f(\xi,\eta)\delta(x-\xi, y-\eta)\,d\xi\,d\eta = f(x,y) \otimes \otimes \delta(x,y) \tag{5.48}$$

If the linear imaging operator S operates on the function object $f(x, y)$ to produce a resultant image function $g(x, y)$ such that

$$g(x, y) = S[f(x, y)] \tag{5.49}$$

then using the linearity property of S and the integral property of a function and a shifted delta function given in Equation 5.4, we have

$$g(x,y) = S\left[\int_{-\infty}^{+\infty}\int_{-\infty}^{+\infty} f(\xi,\eta)\delta(x-\xi, y-\eta)\,d\xi\,d\eta\right] = \int_{-\infty}^{+\infty}\int_{-\infty}^{+\infty} f(\xi,\eta)S[\delta(x-\xi, y-\eta)]\,d\xi\,d\eta \tag{5.50}$$

Note that in Equation 5.50, the term $f(\xi, \eta)$ does not depend explicitly on the variables (x, y) and, therefore, is constant with respect to the operator. However, $S[\delta(x - \xi, y - \eta)]$ is a function of (x, y) and (ξ, η), and this new function is given the function label h (note we are using function labels in alphabetical order here from f to h):

$$S[\delta(x - \xi, y - \eta)] = h(x, y; \xi, \eta) \tag{5.51}$$

The h function can be substituted into (5.50) giving the following equation:

$$g(x,y) = \int\limits_{-\infty}^{+\infty} \int\limits_{-\infty}^{+\infty} f(\xi,\eta)h(x,y;\xi,\eta)\mathrm{d}\xi\mathrm{d}\eta \tag{5.52}$$

The image value g is related to the object transmission f by the *spatial response function h* (Equation 5.52), which can be a function of both "object" (ξ, η) and "image" (x, y) coordinates. To model projection x-ray imaging, $f(\xi, \eta)$ represents the object transmission at the point (ξ, η), while $g(x, y)$ is the resulting image value at the point (x, y). *Note:* $h(x, y; \xi, \eta)$ is better known to imaging physicists and engineers as the *point spread function*. Loosely, $h(x, y; \xi, \eta)$ is the image from an infinitesimal object (i.e., delta function) at the point (ξ, η) in the object plane. In general, h can be a very complicated function of object and image coordinates. The importance of a linear shift-invariant system is that the point spread function h is the same for all locations, greatly simplifying usage. In this case, the spread in h only depends on the difference in coordinates:

$$h(x, y; \xi, \eta) = h(x - \xi, y - \eta) \tag{5.53}$$

Simply stated that if the image of a point has the same functional form as the point is moved around in the object's *x-y* plane (Figure 5.2), then the imaging system is shift invariant. While this is not exactly true for projection x-ray imaging systems, it is approximately true over any sufficiently small area.

Substituting Equation 5.53 into Equation 5.52 gives an equation for $g(x, y)$ expressed as a convolution:

$$g(x,y) = \int\limits_{-\infty}^{+\infty} \int\limits_{-\infty}^{+\infty} f(\xi,\eta)h(x-\xi,y-\eta)\mathrm{d}\xi\mathrm{d}\eta = f(x,y)\otimes\otimes h(x,y) \tag{5.54}$$

This result is important because it shows that an image can be expressed as the convolution of the object with the system point spread function. Equation 5.54 holds regardless of the type of imaging system, making this a very general approach to all forms of imaging systems.

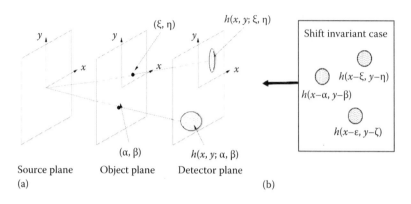

FIGURE 5.2 Geometry of source projection. For the general case, the spread function h changes with location (a). For a shift invariant system, h does not change with location but rather depends only on the difference between output coordinates (b). For a shift invariant imaging system, the image g can be represented as "convolution" between the object f and the point spread function h as follows:

$$g(x,y) = \int\limits_{-\infty}^{+\infty} \int\limits_{-\infty}^{+\infty} f(\xi,\eta)h(x-\xi,y-\eta)d\xi d\eta = f(x,y) \otimes \otimes h(x,y)$$

So how does this mathematical formalism apply to medical imaging (or projection x-ray imaging in particular)? We can think of an object as being comprised of an infinite number of point objects, each with differential area $d\xi d\eta$ where the transmission at each point is equal to the object transmission $f(\xi, \eta)$. We can then image a single point $f(\xi, \eta)$ by fixing (ξ, η) and letting (x, y) vary, producing the point image $dg(x, y)$ as follows:

$$dg(x, y) = f(\xi, \eta)\, h(x - \xi, y - \eta)d\xi\, d\eta \qquad (5.55)$$

Here, $dg(x, y)$ is just an image of the point spread function h modulated by the object transmission at (ξ, η). Repeating by varying (ξ, η) to span the object produces a multiplicity of point spread functions, one for each point on the object that are overlaid and added to each other. Equation 5.54 shows that the sum of all projected point spread images, obtained by integration, gives us the radiographic image g (i.e., photon fluence) produced by the transmission object f.

For a linear shift invariant system, we see that the point spread function h is the key functional component in image formation. Since it measures how object points spread out, it also serves as the basis for assessing the spatial resolution of an imaging system. For a linear space invariant system, we can determine the point spread function $h(x, y)$ by imaging an infinitesimal object (a delta function), and this provides a comprehensive measure of spatial resolution.

The description of the imaging process based on convolution is helpful since we can utilize the property that the Fourier transform of the convolution of two functions equals the product of their Fourier transforms. That is, if $F(u, v)$, $G(u, v)$, and $H(u, v)$ are the

Fourier transforms of $f(x, y)$, $g(x, y)$, and $h(x, y)$, then in the spatial frequency domain, Equation 5.54 becomes

$$G(u, v) = \Im[g(x, y)] = \Im[f(x, y) \otimes \otimes h(x, y)]$$

$$G(u, v) = \Im[f(x, y)]\,\Im[h(x, y)] = F(u, v)\,H(u, v) \tag{5.56}$$

$$\Im\{\text{Image}\} = \Im\{\text{Object}\}\,\Im\{\text{Point Spread Function}\}$$

The Fourier transform simplifies many of our calculations since it converts the convolution of two functions (an integral equation) in the spatial domain into an operation multiplying two functions in the spatial-frequency domain. The product of $F(u, v)$ and $H(u, v)$ is often easier than computing the convolution of $f(x, y)$ with $h(x, y)$ since Fourier transforms can be looked up in standard tables or otherwise determined. The Fourier transform of most functions leads to both real and imaginary parts requiring the use of complex multiplication, but this is more straightforward than spatial domain convolution in many cases.

The Fourier transform also provides insight into the imaging process as a means to represent a system's spatial frequency response. We can describe the response of a system to a simple sinusoidal input. If we know the general response to sinusoids (i.e., at all possible spatial frequencies with corresponding phases), then we know the response of the system to any well-behaved function (i.e., object). Since a point object can be assumed to have a constant magnitude at all frequencies (Equation 5.30), the frequency response (departure from this constant value) of an imaging system given by the Fourier transform of its point spread function is the system transfer function $H(u, v)$. *Since the system transfer function $H(u, v)$ is often a complex function (having nonzero real and imaginary parts), we will concentrate on its magnitude, known as the modulation transfer function or $MTF(u, v)$.* You will become familiar with the MTF in the following chapters.

5.5 GEOMETRY-BASED MODEL OF X-RAY IMAGE FORMATION

In Section 5.4, we used a mathematically elegant approach (borrowed from Macovski) to describe the imaging process in terms of convolution between an object and the point spread function, which implicitly assumed zero magnification. However, projection x-ray imaging is based on geometry and this concept is not obvious from the formalism presented in the previous section.

In this section, we will demonstrate how projection x-ray imaging can be expressed as a convolution, while accounting for its inherent magnification. The approach (borrowed from Barrett and Swindell) will be presented geometrically, which will help to reveal both features and limitations of projection x-ray imaging.

In this derivation, we will use the geometry defined in Figure 5.3 in which the x-ray "source" is located in the $x_s y_s$-plane, the "object" of interest in the $x_o y_o$-plane, and the image

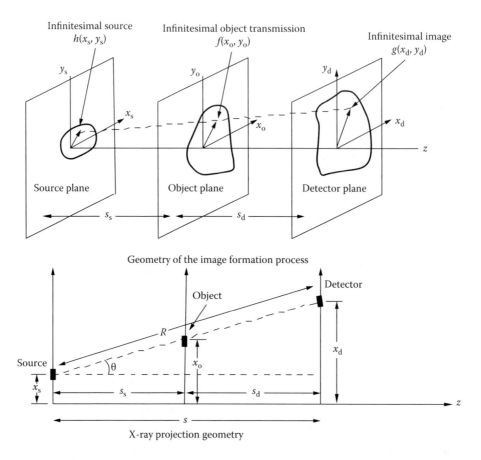

FIGURE 5.3 Geometry and triangles used to show the relationship between variables used in the math of image formation by projection radiography. The x-ray tube focal spot is on the source plane, tissue is on the object plane, and the radiographic image is formed at the detector plane.

in the "detector" $x_d y_d$-plane. The three planes are parallel and share a common z-axis. For this derivation, let us make the following assumptions:

$h(x_s, y_s)$ = emitted photon fluence (photons/area/s) at location (x_s, y_s) in the source plane. The emission is assumed to be isotropic and nonzero only where photons are emitted.

$f(x_o, y_o)$ = x-ray transmission of the object at point (x_o, y_o) in the object plane.

$g(x_d, y_d)$ = incident photon fluence in the detector plane (photons/area/s).

From the definition of photon fluence $g(x_d, y_d)$ [photons/area/s], the number of photons incident on the detector per second in a small area $dx_d dy_d$ at position x_d, y_d is

$$dN_d/dt = g(x_d, y_d)\, dx_d\, dy_d \tag{5.57}$$

The number of photons incident on the detector per second can also be calculated from the source and object geometry as

$$\frac{dN_d}{dt} = \begin{pmatrix} \text{Number of photons} \\ \text{emitted per second} \end{pmatrix} \times \begin{pmatrix} \text{Fraction of photons} \\ \text{incident/area at detector} \end{pmatrix} \times \begin{pmatrix} \text{transmission of} \\ \text{the object} \end{pmatrix} \tag{5.58}$$

$$\frac{dN_d}{dt} = \left(\iint h(x_s, y_s) dx_s \, dy_s \right) \times \left(\frac{\cos(\theta) dx_d dy_d}{4\pi R^2} \right) \times \left(f(x_o, y_o) \right) \tag{5.59}$$

Setting Equation 5.57 equal to Equation 5.59, we obtain the photon fluence g incident at position (x_d, y_d) of the detector:

$$g(x_d, y_d) = \int_{-\infty}^{+\infty} \int_{-\infty}^{+\infty} \frac{\cos\theta}{4\pi R^2} h(x_s, y_s) f(x_o, y_o) dx_s dy_s \tag{5.60}$$

For simplicity, we will assume small angles where

$$\cos\theta \approx 1 \tag{5.61}$$

and

$$R \approx s = s_s + s_d \tag{5.62}$$

where s is the source-to-detector distance. Therefore, Equation 5.60 simplifies as

$$g(x_d, y_d) = \frac{1}{4\pi s^2} \int_{-\infty}^{+\infty} \int_{-\infty}^{+\infty} h(x_s, y_s) f(x_o, y_o) dy_s dx_s \tag{5.63}$$

Equation 5.63 contains coordinates in three different planes (source, object, and detector). If there were no magnification, the detector and object coordinates would be the same, but magnification must be accounted for, so to deal with this, we will express all coordinates at the detector plane. This is important since our measurements are made at the detector plane, and magnification can be used to express corresponding coordinates at the object or source planes. Similar triangles (Figure 5.3) that show that

$$\frac{x_0 - x_s}{s_s} = \frac{x_d - x_0}{s_d} \tag{5.64}$$

where s_s and s_d are the distances from object to source and object to detector planes. Solving Equation 5.64 for x_o leads to an equation for the x-coordinate at the object plane as a function of source and detector plane x-values:

$$x_0 = \frac{s_d}{s_s + s_d} x_s + \frac{s_s}{s_s + s_d} x_d = \left(\frac{M-1}{M}\right) x_s + \frac{1}{M} x_d = \frac{1}{M}\left[(M-1)x_s + x_d\right] \qquad (5.65)$$

Here M is the object magnification (Figure 5.4):

$$M = \frac{s_s + s_d}{s_s} \qquad (5.66)$$

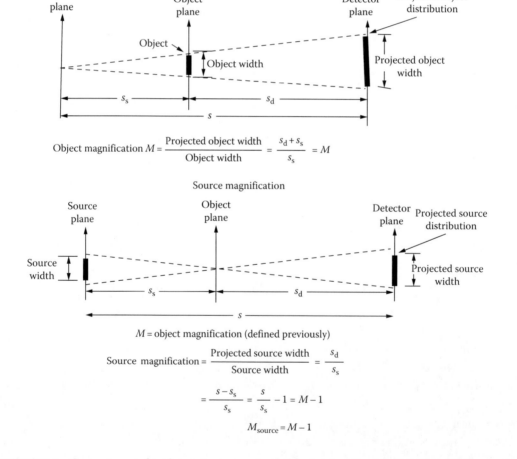

FIGURE 5.4 Geometry used to determine source and object magnification (M). The object magnification (M) is equal to the projected object width divided by the object width. The source magnification ($M - 1$) is equal to the projected source width (through a pinhole placed in the object plane) divided by the source width.

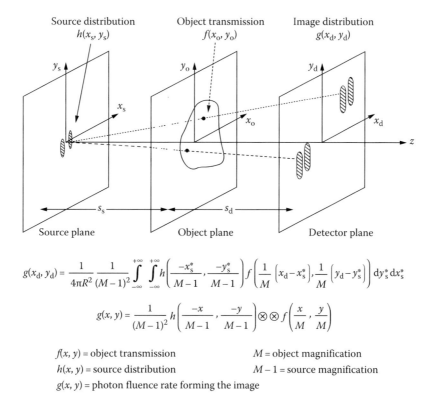

$$g(x_d, y_d) = \frac{1}{4\pi R^2} \frac{1}{(M-1)^2} \int_{-\infty}^{+\infty} \int_{-\infty}^{+\infty} h\left(\frac{-x_s^*}{M-1}, \frac{-y_s^*}{M-1}\right) f\left(\frac{1}{M}\left(x_d - x_s^*\right), \frac{1}{M}\left(y_d - y_s^*\right)\right) dy_s^* dx_s^*$$

$$g(x, y) = \frac{1}{(M-1)^2} h\left(\frac{-x}{M-1}, \frac{-y}{M-1}\right) \otimes \otimes f\left(\frac{x}{M}, \frac{y}{M}\right)$$

$f(x, y)$ = object transmission M = object magnification

$h(x, y)$ = source distribution $M - 1$ = source magnification

$g(x, y)$ = photon fluence rate forming the image

FIGURE 5.5   The source distribution h produces a magnified image distribution g for each point in the object. Integrating over all contributions from the source and object forms the final image. Here, the focal spot (source) is modeled as a bimodal distribution.

Similarly, the equation for the y-coordinate is

$$y_0 = \frac{s_d}{s_s + s_d} y_s + \frac{s_s}{s_s + s_d} y_d = \left(\frac{M-1}{M}\right) y_s + \frac{1}{M} y_d = \frac{1}{M}\left[(M-1)y_s + y_d\right] \qquad (5.67)$$

Substituting x_0 and y_0 from Equations 5.65 and 5.67 into (5.63) yields

$$g(x_d, y_d) = \frac{1}{4\pi s^2} \int_{-\infty}^{+\infty} \left[\int_{-\infty}^{+\infty} h(x_s, y_s) f\left(\frac{1}{M}\left((M-1)x_s + x_d\right), \frac{1}{M}\left((M-1)y_s + y_d\right)\right) dy_s \right] dx_s$$

$$(5.68)$$

Finally, if we project the source coordinates onto the image (detector) plane through a point in the object,

$$x_s^* = -(M-1)x_s \quad \text{and} \quad y_s^* = -(M-1)y_s \qquad (5.69)$$

Equation 5.68 can now be expressed in terms of coordinates at the detector plane based on source magnification ($M - 1$) and object magnification (M):

$$g(x_d, y_d) = \frac{1}{4\pi s^2} \frac{1}{(M-1)^2} \int\limits_{-\infty}^{+\infty} \left[\int\limits_{-\infty}^{+\infty} h\left(\frac{-x_s^*}{M-1}, \frac{-y_s^*}{M-1}\right) f\left(\frac{1}{M}(x_d - x_s^*), \frac{1}{M}(y_d - y_s^*)\right) dy_s^* \right] dx_s^*$$

(5.70)

On the right side of Equation 5.70, all coordinates indicated with "*" are in the detector plane, so the asterisk and subscripts can be dropped. This equation is then seen to be the convolution of the source distribution h with the object transmission distribution f where magnification is included:

$$g(x,y) = \frac{1}{(M-1)^2} h\left(\frac{-x}{M-1}, \frac{-y}{M-1}\right) \otimes \otimes f\left(\frac{x}{M}, \frac{y}{M}\right)$$

(5.71)

The subscripts were removed without loss of mathematical generality (Figure 5.5), and we have chosen the distance units of R so that $4\pi s^2 = 1$. One can calculate appropriate scale factors to account for such issues by making measurements of h with a pinhole camera. *The interpretation of Equation 5.71 is that for projection x-ray imaging, the image "g" is formed as the convolution of the magnified source "h" with the magnified object "f". This result is consistent with the general imaging equation (Equation 5.54). This section, based on the geometry of projection x-ray imaging, shows that the "magnified source is the 'point spread function'" presented in the prior section!*

It is instructive to look at Equation 5.71 for two special cases.

CASE 5.1:   IMAGING A POINT OBJECT (FIGURE 5.6)

If we have a point object $f(x/M, y/M) = \delta(x/M)\,\delta(y/M)$, then the image g is equal to

$$g(x,y) = \left(\frac{M}{M-1}\right)^2 h\left(\frac{-x}{M-1}, \frac{-y}{M-1}\right)$$

(5.72)

the source h magnified by the factor $M - 1$. Indeed, we see that if we image a point object with a magnification M, the source distribution h is projected onto the detector with a magnification $M - 1$. In this case, the image is a magnified version of the source. So, if we image a point object (i.e., a pinhole camera), we obtain an image of the source distribution magnified by a factor of $M - 1$. The minus sign in the arguments of h provides the spatial reversal when the source is projected through a point object.

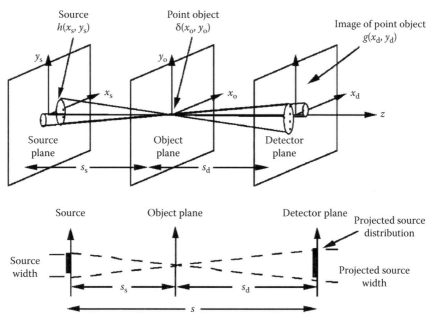

$M - 1$ = source magnification = projected source width/source width = s_d/s_s

Imaging equation: $\qquad g(x,y) = \dfrac{1}{(M-1)^2}\, h\left(\dfrac{-x}{M-1}, \dfrac{-y}{M-1}\right) \otimes \otimes f\left(\dfrac{x}{M}, \dfrac{y}{M}\right)$

Special case 1: *Imaging a point object*

If $f\left(\dfrac{x}{M}, \dfrac{y}{M}\right) = \delta\left(\dfrac{x}{M}\right)\delta\left(\dfrac{y}{M}\right) = M^2\,\delta(x)\,\delta(y)$ (point object), then

$$g(x,y) = \left(\dfrac{M}{(M-1)}\right)^2 h\left(\dfrac{-x}{(M-1)}, \dfrac{-y}{(M-1)}\right) \qquad \text{(magnified source)}$$

FIGURE 5.6 In the limiting case of a point object (such as for a pinhole camera), the resulting image is a magnified and spatially reversed version of the source.

CASE 5.2: IMAGING WITH A POINT SOURCE (FIGURE 5.7)

If we now image with a point source $h(x, y) = \delta(x/(M-1))\,\delta(y/(M-1))$, then Equation 5.71 becomes

$$g(x,y) = f\left(\dfrac{x}{M}, \dfrac{y}{M}\right) \tag{5.73}$$

the object magnified by M. Note that the negative argument can be made positive since the delta function only exists at $x, y = 0, 0$. Again, this is precisely what we expect, and is highly desirable for projection radiography. As shown in Figure 5.7, with an ideal (point) source, the object is projected onto the image receptor but is magnified by the factor M. These two cases are good examples of why M is "object magnification" and $M - 1$ is "source magnification."

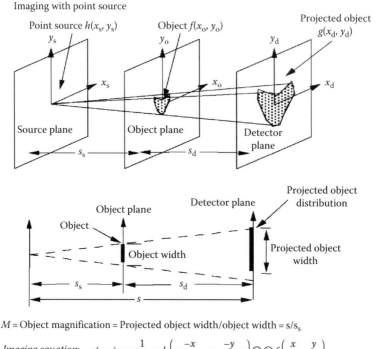

M = Object magnification = Projected object width/object width = s/s_s

Imaging equation: $g(x, y) = \dfrac{1}{(M-1)^2} h\left(\dfrac{-x}{M-1}, \dfrac{-y}{M-1}\right) \otimes \otimes f\left(\dfrac{x}{M}, \dfrac{y}{M}\right)$

Special case 2: Imaging a point source

If $h\left(\dfrac{-x}{M-1}, \dfrac{-y}{M-1}\right) = \delta\left(\dfrac{-x}{M-1}\right)\delta\left(\dfrac{-y}{M-1}\right) = (M-1)^2 \delta(x)\delta(y)$ a (point object), then

$g(x, y) = f\left(\dfrac{x}{M}, \dfrac{y}{M}\right)$ (magnified object)

FIGURE 5.7   In the limiting case of a point source, the image is a magnified version of the object.

In reality, we have a small x-ray source $h(x, y)$ (focal spot) irradiating an extended object $f(x, y)$. One way to view imaging is to consider the object as an infinite collection of points. Each small point in the object projects an image of the extended source (focal spot) onto the image plane with corresponding attenuation. Finally, the net image is equal to the sum of these projected focal spot images.

Alternatively, we can treat the extended source as a collection of point sources. Each point source projects an image of the object onto the image receptor. The final image is obtained by summing the projected images from each of the points comprising the extended source.

Mathematically, the image forming operation is quite straightforward if we know the source distribution $h(x, y)$ and the object transmission $f(x, y)$. If the object is magnified by a factor of M onto the detector, then the source h is magnified by a factor of $M - 1$. These magnified distributions are given by

$$\text{Magnified source} = \dfrac{1}{(M-1)^2} h\left(\dfrac{-x}{M-1}, \dfrac{-y}{M-1}\right)$$

$$\text{Magnified object} = f\left(\frac{x}{M}, \frac{y}{M}\right)$$

Importantly, if we evaluate these when $M = 1$, we see that the magnified source becomes a delta function and that the object is not magnified.

5.6 FREQUENCY DOMAIN MODEL OF IMAGE FORMATION

The imaging process can be represented in the spatial-frequency domain by taking the Fourier transform of Equation 5.71. If we maintain our notation where (u, v) are the spatial frequency variables corresponding to the spatial variables (x, y), and if the functions f, g, and h have Fourier transforms F, G, and H, respectively, then taking the Fourier transform of Equation 5.71 gives

$$G(u, v) = M^2 F(Mu, Mv)\,[H(-(M-1)u, -(M-1)v)] \tag{5.74}$$

Note that when $M = 1$, then $H(u, v) = H(0, 0) = \text{constant}$, with magnitude $= 1$ if the integral of $h(x, y) = 1$ (i.e., normalized to unit area); therefore, $G(u, v) = F(u, v)$, as expected. Hence, in the spatial frequency domain, the Fourier transform of the image G is obtained by multiplying the Fourier transform of the magnified source distribution H by the Fourier transform of the magnified object transmission F. As described before, h is the point spread function and H the system transfer function.

The representation of an imaging system in the frequency domain also provides a useful tool in the analysis of image systems with multiple stages. If the image of an object with transmission f is formed through a sequence of steps, where each step is a linear spatially invariant system represented by point spread functions h_1, h_2, ..., h_n, then the image g is given by multiple convolutions:

$$g = f \otimes h_1 \otimes h_2 \otimes \ldots \otimes h_n \tag{5.75}$$

In the frequency domain, the product of the Fourier transforms of these spread functions leads to a frequency domain image:

$$G = F \cdot (H_1 H_2 \ldots H_n) \tag{5.76}$$

so that the composite system transfer function H-system is given by the product of the transfer functions of each stage:

$$H_{\text{system}} = H_1 H_2 \ldots H_n \tag{5.77}$$

Example 5.1

Emission of x-rays from a realistic focal spot is not uniform in the direction perpendicular to the anode–cathode axis. Rather, in 1-D, it has the form of a double Gaussian (Figure 5.8):

$$h(x) = e^{\frac{-\pi(x-a)^2}{b^2}} + e^{\frac{-\pi(x+a)^2}{b^2}} \tag{5.78}$$

We will use this focal-spot distribution function to illustrate several mathematical concepts.

(a) Prove that the convolution of two Gaussian functions is another Gaussian function. Specifically, show that

$$\frac{1}{b}e^{\frac{-\pi x^2}{b^2}} \otimes \frac{1}{c}e^{\frac{-\pi x^2}{c^2}} = \frac{1}{\sqrt{b^2+c^2}}e^{\frac{-\pi x^2}{b^2+c^2}} \tag{5.79}$$

(b) Explain why $h(x)$ in (5.78) can be expressed in terms of a convolution with two delta functions:

$$h(x) = e^{\frac{-\pi x^2}{b^2}} \otimes \left[\delta(x+a) + \delta(x-a) \right] \tag{5.80}$$

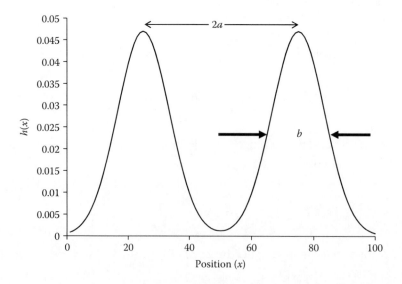

FIGURE 5.8 In one dimension, a more realistic focal spot $h(x)$ can be represented as the sum of two Gaussian functions having separation $2a$ and width b.

(c) Use (a) and (b) to show that if this focal spot distribution is used to image a 1-D object with a Gaussian transmission function

$$f(x) = e^{\frac{-\pi x^2}{c^2}} \tag{5.81}$$

and an object magnification M, then the image is represented by the function

$$g(x) = \frac{bcM}{\sqrt{(M-1)^2 b^2 + M^2 c^2}} \left\{ \exp\left[\frac{-\pi \left[x - (M-1)a\right]^2}{(M-1)^2 b^2 + M^2 c^2} \right] \right.$$

$$\left. + \exp\left[\frac{-\pi \left[x + (M-1)a\right]^2}{(M-1)^2 b^2 + M^2 c^2} \right] \right\} \tag{5.82}$$

(d) Interpret the result obtained in (c) geometrically.

Solution

(a) We can prove Equation 5.79 in the spatial frequency domain. Making use of the properties of Fourier transforms and Equation 5.35, we have

$$\Im\left(e^{\frac{-\pi x^2}{b^2}} \otimes e^{\frac{-\pi x^2}{c^2}} \right) = \Im\left(e^{\frac{-\pi x^2}{b^2}} \right) \Im\left(e^{\frac{-\pi x^2}{c^2}} \right)$$

$$\Im\left(e^{\frac{-\pi x^2}{b^2}} \right) \Im\left(e^{\frac{-\pi x^2}{c^2}} \right) = \left(b\, e^{-\pi b^2 u^2} \right)\left(c\, e^{-\pi c^2 u^2} \right) = bc\, e^{-\pi\left(b^2 + c^2\right)u^2}$$

$$bce^{-\pi\left(b^2 + c^2\right)u^2} = \Im\left(\frac{bc}{\sqrt{b^2 + c^2}}\, e^{\frac{-\pi x^2}{b^2 + c^2}} \right)$$

$$\Im\left(e^{\frac{-\pi x^2}{b^2}} \otimes e^{\frac{-\pi x^2}{c^2}} \right) = \Im\left(\frac{bc}{\sqrt{b^2 + c^2}}\, e^{\frac{-\pi x^2}{b^2 + c^2}} \right) \tag{5.83}$$

such that

$$e^{\frac{-\pi x^2}{b^2}} \otimes e^{\frac{-\pi x^2}{c^2}} = \frac{bc}{\sqrt{b^2+c^2}} e^{\frac{-\pi x^2}{b^2+c^2}} \tag{5.84}$$

(b) From Equation 5.13, we have for the focal spot distribution $h(x)$

$$h(x) = e^{\frac{\pi x^2}{b^2}} \otimes [\delta(x+a)+\delta(x-a)] = \left[e^{\frac{\pi x^2}{b^2}} \otimes \delta(x+a) \right] + \left[e^{\frac{\pi x^2}{b^2}} \otimes \delta(x-a) \right] \tag{5.85}$$

such that

$$h(x) = e^{\frac{\pi(x-a)^2}{b^2}} + e^{\frac{-\pi(x+a)^2}{b^2}}$$

(c) From Equation 5.71, the 1-D image distribution $g(x)$ is seen as the convolution of the magnified focal spot distribution h (point spread function) with the magnified object distribution f:

$$g(x) = \frac{1}{M-1} f\left(\frac{x}{M}\right) \otimes h\left(\frac{-x}{M-1}\right) \tag{5.86}$$

After substitution using Equations 5.80 and 5.81, we have

$$g(x) = \frac{1}{M-1} \exp\left[\frac{-\pi x^2}{M^2 c^2}\right] \otimes \left\{ \exp\left[\frac{-\pi x^2}{(M-1)^2 b^2}\right] \otimes \left[\delta\left(\frac{-x}{M-1}+a\right)+\delta\left(\frac{-x}{M-1}-a\right)\right]\right\} \tag{5.87}$$

and using the associative property of convolution (Equation 5.11),

$$g(x) = \left\{ \frac{1}{M-1} \exp\left[\frac{-\pi x^2}{M^2 c^2}\right] \otimes \exp\left[\frac{-\pi x^2}{(M-1)^2 b^2}\right] \right\} \otimes \left[\delta\left(\frac{-x}{M-1}+a\right)+\delta\left(\frac{-x}{M-1}-a\right)\right] \tag{5.88}$$

We can compute the convolution of two Gaussian functions using the result of part (a) of this example

$$g(x) = \frac{bcM}{\sqrt{(M-1)^2 b^2 + M^2 c^2}} \left\{ \exp\left[\frac{-\pi x^2}{(M-1)^2 b^2 + M^2 c^2} \right] \right\}$$

$$\otimes \left[\delta\left(\frac{-x + (M-1)a}{M-1} \right) + \delta\left(\frac{-x - (M-1)a}{M-1} \right) \right] \qquad (5.89)$$

And evaluating the convolution leads to the following:

$$g(x) = \frac{bcM}{\sqrt{(M-1)^2 b^2 + M^2 c^2}} \left\{ \exp\left[\frac{-\pi\left[x - (M-1)a\right]^2}{(M-1)^2 b^2 + M^2 c^2} \right] + \exp\left[\frac{-\pi\left[x + (M-1)a\right]^2}{(M-1)^2 b^2 + M^2 c^2} \right] \right\}$$

$$(5.90)$$

(d) A helpful geometrical interpretation of the image can be seen in Equations 5.89 and 5.90. The two delta functions determine the positions in the image plane for the two lobes of the focal spot. The distance between these two lobes is $2(M-1)a$. The width of the lobes is in the Gaussian (exponential term) and based on both the width of the magnified object (Mc) and the width of the magnified focal spot $(M-1)b$.

If the magnified width of the Gaussian terms (lobes) is much larger than the distance between the two images (i.e., if $\sqrt{(M-1)^2 b^2 + M^2 c^2} \gg 2(M-1)a$), then the two Gaussian distributions will blur together and appear as one. If the magnified width of the Gaussian term is much smaller than the distance between the two images (i.e., if $\sqrt{(M-1)^2 b^2 + M^2 c^2} \ll 2(M-1)a$), then two Gaussian distributions can be distinguished. Additionally if the width of the magnified object (Mc) is much larger than that of the magnified focal spot distributions $(M-1)b$, then Equation 5.90 simplifies to

$$g(x) = b\left\{ \exp\left[\frac{-\pi\left[x - (M-1)a\right]^2}{M^2 c^2} \right] + \exp\left[\frac{-\pi\left[x + (M-1)a\right]^2}{M^2 c^2} \right] \right\} \qquad (5.91)$$

with two distinct images of the object Gaussian lobes.

Most importantly as M approached unity, which is the case for many radiographic procedures, Equation 5.91 reduces to

$$g(x) = 2be^{\frac{-\pi x^2}{c^2}} \qquad (5.92)$$

In this case, both the effect of the focal spot width and its two lobes are eliminated and the image appearance is that of the object alone.

5A APPENDIX

Scaling and shifting of common imaging functions

An integrable function $f(x)$ can be height and width scaled and shifted along x.

$$a \cdot f\left(\frac{x-b}{c}\right) \begin{cases} a = \text{height scale} \\[6pt] b = \text{origin shift} \\[6pt] c = \text{width scale} \\[6pt] \displaystyle\int_{-\infty}^{+\infty} a \cdot f\left(\frac{x-b}{c}\right)dx = a \cdot c \int_{-\infty}^{+\infty} f(x)dx \end{cases}$$

The integral of the height and width scaled function is the integral of the original function multiplied by the product of the height and width scales $a \cdot c$. We refer to $a \cdot c$ as the area scale, and if $a = 1/c$, then the area is unchanged by the scaling. The shift term does not affect area as long as the limits of integration span the range of the function.

Scaling and shifting of template functions

The normalized sinc function is

$$\sin c(x) = \frac{\sin(\pi x)}{\pi x} \begin{cases} = 1 & \text{at } x = 0 \\[6pt] = 0 & \text{at } x = \pm n \quad \text{for } n > 0 \text{ and } n \text{ is an integer.} \\[6pt] \displaystyle\int_{-\infty}^{+\infty} \sin c(x)dx = 1 \end{cases}$$

The scaled and shifted sinc function is then

$$a \cdot \sin c\left(\frac{x-b}{c}\right) = a \cdot \frac{\sin\left[\pi\left(\frac{x-b}{c}\right)\right]}{\pi\left(\frac{x-b}{c}\right)} \begin{cases} = a & \text{at } x = b \\[6pt] = 0 & \text{at } \dfrac{x-b}{c} = \pm n \quad \text{for } n > 0 \quad \text{and } n \text{ is an integer.} \\[6pt] \displaystyle\int_{-\infty}^{+\infty} a \cdot \sin c\left(\frac{x-b}{c}\right)dx = a \cdot c \int_{-\infty}^{+\infty} \sin c(x)dx = a \cdot c \end{cases}$$

Note: If $a = 1/c$, then the area for the scaled sinc function remains unity.

The scaling and shifting properties for other normalized template functions (rectangle, triangle, and Gaussian) follow the trend given for the sinc function.

Relationship between rect and sinc functions

Several interesting results are seen when we evaluate the Fourier transform of $f(x) = \text{rect}(x/a)$.

1. $\Im[f(x)] = \Im\left[\text{rect}\left(\dfrac{x}{a}\right)\right] = \displaystyle\int_{-\frac{a}{2}}^{+\frac{a}{2}} e^{-2\pi iux}\,dx$

2. $= \dfrac{e^{-2\pi iu\frac{a}{2}} - e^{2\pi iu\frac{a}{2}}}{-2\pi ui} = \left(\dfrac{e^{\pi iua} - e^{-\pi iua}}{2i}\right)\dfrac{1}{\pi u} = \dfrac{\sin(\pi ua)}{\pi u} = a\dfrac{\sin(\pi ua)}{\pi ua}$

3. so $\Im[f(x)] = F(u) = a\sin c(au)$

From this simple exercise, we see that $f(x) = \text{rect}(x/a)$, with width scale $= a$, transforms to

$$F(u) = a \cdot \sin c(au) \text{ with height scale} = a \text{ and width scale of } 1/a.$$

Result 1. Since the height of the rectangle $= 1$, then $f(0) = 1$ and the integral of $a \cdot \sin c(au) = 1$. This is easily confirmed since the integral of $\sin c(u)$ is unity and the area scale of $a \cdot \sin c(au)$ is unity.

Result 2. Since $F(0) = a$, the integral of $\text{rect}(x/a) = a$. This is easily confirmed for $\text{rect}(x/a)$ since its height $= 1$ and its width $= a$.

Result 3. Setting $a = 1$ leads to the classic unit area templates, $f(x) = \text{rect}(x)$ and $F(u) = \sin c(u)$, as Fourier transform pairs.

Result 4. For the spatial domain function $\text{rect}(x/a)$, the corresponding frequency domain $a \cdot \sin c(au)$ first goes to zero at $u = 1/a$. This supports a sinc function that goes to zero at frequencies that are noninteger multiples of π. These zero points are independent of height scaling.

Result 5. In the limit as a goes to infinity, $\text{rect}(x/a)$ approaches a unity valued constant over all x. We know that the Fourier transform of such a constant is $\delta(u)$ leading to the following:

$$\delta(u) \Rightarrow \Im\left[\text{rect}\left(\dfrac{x}{a}\right)\right] = \int_{-\frac{a}{2}}^{+\frac{a}{2}} e^{-2\pi iux}\,dx$$

$$\text{limit } a \Rightarrow \infty$$

HOMEWORK PROBLEMS

P5.1 We sometimes get confused about the correct units for the spatial frequency variable u, in particular, whether u is expressed in terms of cycles/length, radians/length, or degrees/length (or some other unit), all of which would be reasonable units for a spatial frequency variable.

(a) Please answer this question: If the spatial variable x has units of mm, what are the units of the conjugate spatial frequency variable u?

(b) We know that if $\Im$ denotes the Fourier transform operator, then

$$\Im\left[\cos\left(\frac{\pi x}{a}\right)\right] = \frac{1}{2}\left[\delta\left(u - \frac{1}{2a}\right) + \delta\left(u + \frac{1}{2a}\right)\right]$$

First, graph the function

$$f(x) = \cos\left(\frac{\pi x}{a}\right)$$

and label your axes carefully. Use your graph to determine the spatial frequency of $f(x)$ in the appropriate units. Then graph

$$F(u) = \frac{1}{2}\left[\delta\left(u - \frac{1}{2a}\right) + \delta\left(u + \frac{1}{2a}\right)\right]$$

and explain how the location of the delta functions comprising $F(u)$ substantiates your answer in part (a) with respect to the correct units for the spatial frequency variable u.

(c) As long as we are discussing units, what are the units of $\delta(x)$?

Big hint: This question is easy to answer if you remember that

$$\int_{-\infty}^{+\infty} \delta(x)dx = 1$$

P5.2 The following problems demonstrate the versatility of the Fourier transform.

(a) Use the definition of the Fourier transform to prove that if $F(u)$ is the Fourier transform of a function $f(x)$, then

$$\int_{-\infty}^{+\infty} f(x)dx = F(0)$$

(b) Use the result from (a) to show that

$$\int\limits_{-\infty}^{+\infty} \frac{\sin(ax)}{x} dx = \pi$$

P5.3 A function is symmetric about the origin if $f(x) = f(-x)$ for all values of x. Show that if a function is symmetric, then its Fourier transform is real (i.e., contains no imaginary components).

P5.4 Prove using direct application of the Fourier transform equations and any of its properties (5.30, 5.31, 5.34, and 5.35).

CHAPTER 6

Spatial Resolution

6.1 DEFINITIONS AND MEASUREMENT OF SPATIAL RESOLUTION

6.1.1 Point Spread Function

In the spatial domain, the resolution of an imaging system is characterized by its point spread function (psf) (Figure 6.1). For 2-D images, the psf(x, y) is the image obtained of an infinitesimal point object that can be defined as the product of two delta functions:

$$\text{point}(x,y) = \delta(x,y) = \delta(x)\delta(y) \tag{6.1}$$

If the imaging system is represented mathematically by the transform S, then the system point spread function psf(x, y) is obtained by transforming the point object (Figure 6.1):

$$\text{psf}(x,y) = S\left[\text{point}(x,y)\right] \tag{6.2}$$

If (u, v) are the conjugate spatial frequency variables for the spatial variables (x, y), then the modulation transfer function MTF(u, v) is obtained from the point spread function psf(x, y) as the magnitude of its two-dimensional Fourier transform:

$$\text{STF}(u,v) = \Im\{\text{psf}(x,y)\} \tag{6.3}$$

$$\text{MTF}(u,v) = |\text{STF}(u,v)| \tag{6.4}$$

Here, STF(u, v) is the "system transfer function," which is a complex function (having real and imaginary parts or alternatively using magnitude and phase). Since we are often just concerned with the magnitude of a system's response as a function of frequency, the

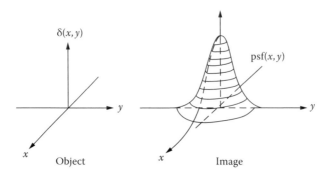

FIGURE 6.1   The inherent blurring of an ideal point $\delta(x, y)$ by the imaging system can be assessed by evaluation of the system's point spread function psf(x, y).

modulation transfer function, or MTF(u, v), is commonly used to assess resolution. *The most important theoretical descriptors of the spatial resolution of a medical imaging system are the psf and the MTF.*

The psf for an x-ray imaging system is difficult to determine in practice because an infinitesimal point object (Figure 6.1) can only be approximated for an x-ray imaging system. A tiny aperture in a radiopaque plate is used to approximate a point object. Since the aperture must be small in comparison with the spatial resolution of the system, very few x-rays are transmitted through the aperture requiring lengthy exposure times. Per specs from one system a 0.010 mm pinhole diameter is recommended for focal spot sizes from 0.5 to 0.10 mm, a 0.030 mm diameter for sizes from 0.10 to 1.0 mm, a 0.075 mm diameter for focal spots from 1.0 to 2.5 mm, and 0.100 mm diameter for sizes above 2.5 mm. After the film image is obtained, a scanning microdensitometer is used to record a profile through the center of the psf. An H&D transform is then applied to convert film density to relative exposure (required for linearity purposes). Linearity is required for all spatial resolution measurements and linearization will be assumed to be a part of the measurement process when the psf for x-ray imaging is recorded using film.

6.1.2 Line Spread Function

Measuring the line spread function (lsf) (Figure 6.2) can reduce technical difficulties associated with obtaining and measuring the psf. As its name suggests, the lsf is obtained with a narrow slit in a radiopaque object, rather than a tiny point aperture. The lsf is an oriented one-dimensional representation of the two-dimensional psf, so images are usually acquired with the line oriented both parallel and perpendicular to the x-ray tube's anode–cathode direction, the two most important directions. Also, the lsf is acquired near the center of the field of view. We record the lsf with a microdensitomer by scanning across the film image of the slit, perpendicular to its length. As for psf determination, the width of the lsf's slit must be sufficiently narrow so that its finite extent does not contribute significantly to the width of the slit's image. That is, the spread in the imaged slit must be due almost entirely to the blurring contributed by the imaging system rather than the width of the slit. For x-ray film-screen systems, a slit width of 0.010 mm is used for this measurement.

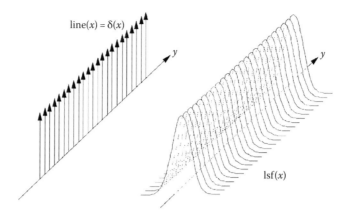

FIGURE 6.2 The line spread function lsf(x) is the image of an ideal line object (e.g., small slit in a lead plate). A 2-D line can be modeled as a 1-D delta function δ(x).

A line (or slit) is defined mathematically as

$$\text{line}(x) = \delta(x) = \int_{-\infty}^{+\infty} \delta(x)\delta(y)\,dy = \int_{-\infty}^{+\infty} \text{point}(x,y)\,dy \tag{6.5}$$

The line is narrow in the x-direction and not a function of the y-coordinate. If S is the linear transform for the imaging system, then the line spread function lsf(x) is represented mathematically as follows:

$$\text{lsf}(x) = S\big[\text{line}(x)\big] = S\left[\int_{-\infty}^{+\infty} \text{point}(x,y)\,dy\right] = \int_{-\infty}^{+\infty} S\big[\text{point}(x,y)\,dy\big] = \int_{-\infty}^{+\infty} \big[\text{psf}(x,y)\big]\,dy \tag{6.6}$$

Therefore, the line spread function lsf(x) is equivalent to the integral along y of the 2-D point spread function psf(x, y). *Note: We are continuing to use capital letters for functions in the frequency domain and lower case letters for functions in the spatial domain.*

The 1-D MTF is the Fourier transform of the lsf. This is shown mathematically as follows:

$$\Im\{\text{lsf}(x)\} = \int_{-\infty}^{+\infty} \text{lsf}(x)e^{-2\pi iux}\,dx = \int_{-\infty}^{+\infty}\left[\int_{-\infty}^{+\infty} \text{psf}(x,y)\,dy\right]e^{-2\pi iux}\,dx \tag{6.7}$$

$$= \int_{-\infty}^{+\infty}\int_{-\infty}^{+\infty} \text{psf}(x,y)e^{-2\pi i(ux+0y)}\,dx\,dy = \Im\{\text{psf}(x,y)\}_{v=0} = \text{STF}(u,0) \tag{6.8}$$

Hence, the Fourier transform of the lsf is the STF evaluated along one dimension (recall that MTF = |STF|). If the psf is spatially symmetric, then STF(u, 0) completely specifies the MTF. The advantage of the lsf over the psf is that we can average multiple samples of the lsf profile at different y-locations to reduce noise, so that less exposure is needed. However, care must be taken to ensure vertical alignment of the lsf such that the peak is at the same location in each scanned profile.

It is possible to remove the effect of finite width of a slit on the measured MTF$(u)^*$ using a process called deconvolution. The measured MTF$(u)^*$ is the product the system's MTF(u) and the magnitude of the slit's Fourier transform, Slit(u). Therefore, the algebraic division MTF$(u)^*$/Slit(u) can then be used to estimate the corrected MTF(u). However, care must be taken where Slit(u) goes to zero, since MTF should also be zero, and in general, the deconvolution (division here) is stopped before the first zero in Slit(u). By using a very narrow slit, this first zero should be well beyond the useful frequency range of the system MTF. This approach to remove the size of the aperture can be applied to the MTF determined from a psf as well.

6.1.3 Edge Spread Function

The final method to measure spatial resolution of an imaging system is based on the "edge spread" function (esf) (Figure 6.3). For this technique, the source is presented with an object that transmits radiation on one side of an edge, which is perfectly attenuating on the other. The transmission for such an edge can be defined mathematically using a unit step function:

$$\text{step}(x,y) = \begin{cases} 1 & \text{if } x \geq 0 \\ 0 & \text{if } x < 0. \end{cases} \tag{6.9}$$

As Barrett and Swindell point out, this step function can also be written as

$$\text{step}(x,y) = \text{step}(x) = \int_{-\infty}^{x} \delta(x')dx' = \int_{-\infty}^{x} \text{line}(x')dx' \tag{6.10}$$

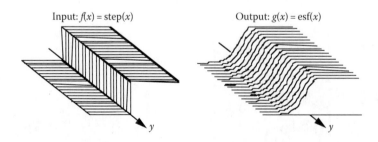

Input: $f(x) = \text{step}(x)$ Output: $g(x) = \text{esf}(x)$

FIGURE 6.3 The edge spread function esf(x) is the image of an ideal step object (e.g., edge of a lead plate). The lsf can be derived as the spatial derivative of the esf. Similarly, a line object is modeled as the derivative of an edge object.

based on the definition of line(x) from Equation 6.5. If the system transform S is linear, then

$$\text{esf}(x) = S\{\text{step}(x)\} = S\left\{\int_{-\infty}^{x} \text{line}(x')dx'\right\} = \int_{-\infty}^{x} S\{\text{line}(x')\}dx' = \int_{-\infty}^{x} \text{lsf}(x')dx' \quad (6.11)$$

where it can be seen that the lsf is just the derivative of the esf:

$$\text{lsf}(x) = \frac{d}{dx}\left[\text{esf}(x)\right] \quad (6.12)$$

A profile curve across the edge image is used as the esf. The derivative of the esf is the lsf, the Fourier transform of which yields the MTF in one direction. The advantage of the esf over the psf- or lsf-based approaches is that the esf can be acquired with a much smaller radiation exposure. As with the lsf, we can average many edge profiles, but we must ensure alignment of the individual edge profiles.

6.2 COMPONENTS OF UNSHARPNESS

Now that we have described system blurring or "unsharpness" mathematically in terms of spread functions, we will use this approach to further study and/or quantify the four major imaging system components of unsharpness in radiographic imaging systems. *Geometric unsharpness* refers to the loss of detail with increasing size of the radiation source (focal spot). *Motion unsharpness* refers to the loss of detail due to motion of the source (focal spot), the detector, or the object being imaged (patient) while the image is being acquired. *Detector unsharpness* refers to the loss of detail due to the resolving power of the detector. Finally, *digitization unsharpness* refers to the loss of detail associated with the conversion of the image from its natural analog form to a digital format, that is, analog-to-digital conversion.

6.2.1 Geometric Unsharpness

Geometric unsharpness in a projection radiograph refers to the loss in image detail caused by the size of the x-ray source, that is, the focal spot. As shown in Figure 6.4, an extended focal spot can blur the appearance of an object. We use language borrowed from astronomy to describe the extent of this blurriness. The area directly behind the object, which is completely contained within the shadow of the object, is called the "umbra." Similarly, the region that has a partial shadow is called the "penumbra." In some texts, the penumbra is called the edge gradient. (*Note: umbra* is Latin for shadow and *paene* is Latin for almost. "Umbrella" has similar Latin roots as *ombrella*.)

Obviously, as the size the focal spot increases, the size of the umbra decreases while that of the penumbra increases, and the degree of geometrical unsharpness or edge blurring increases. Therefore, to obtain the most detail in the image, one should use the smallest

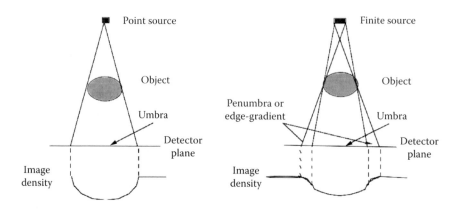

FIGURE 6.4 The geometric component of unsharpness in a projection radiograph is the loss of detail due to the finite size of the x-ray tube focal spot. Increasing focal spot size reduces the umbra (shadow) region and increases the penumbra, leading to blurring at edges.

focal spot possible. However, the focal spot must be large enough to handle the localized heat generated on the anode. This fundamental limitation forces many diagnostic medical procedures to utilize x-ray tubes with focal spots at least 1 mm wide.

Another important factor affecting geometric unsharpness is object magnification M. As we found in Chapter 5, if the object is projected onto the image receptor with a magnification M, the focal spot is magnified by a factor of $M - 1$. When the object is halfway between the source and detector, it is imaged with a magnification of 2, while the projected focal spot size is equal to its physical size (magnification equal to $M - 1 = 1$). As the object is moved closer to the detector, the object magnification tends toward unity, while source or focal spot magnification tends toward zero. As the object is brought closer to the source, the magnification of both the object and the focal spot magnification increases, causing increased geometric unsharpness (Figure 6.5). The increase in geometric unsharpness that accompanies magnification limits the degree to which magnification can be used to increase the spatial resolution with which the object is imaged.

Geometric unsharpness is modeled mathematically using linear systems analysis presented previously in Equation 5.69. In this formulation an image $g(x, y)$ is obtained by convolving the magnified object $f(x/M, y/M)$ with the magnified focal spot distribution $h(-x/(M - 1), -y/(M - 1)$, that is, the magnified psf. This is expressed mathematically as

$$g(x,y) = \frac{1}{(M-1)^2} f\left(\frac{x}{M}, \frac{y}{M}\right) \otimes \otimes h\left(\frac{-x}{M-1}, \frac{-y}{M-1}\right) \tag{6.13}$$

or in the frequency domain

$$G(u,v) M^2 F(Mu, Mv) H(-(M-1)u, -(M-1)v). \tag{6.14}$$

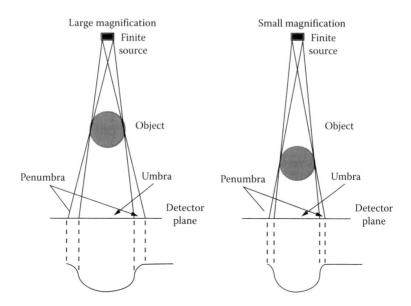

FIGURE 6.5 Reducing the distance from the object to the detector reduces geometric unsharpness.

$H(u, v)$ is the two-dimensional transfer function associated with blurring by the focal spot. Both $f(x, y)$ and $F(u, v)$ refer to the object distributions (in spatial and frequency domains), which are magnified by M onto the plane of the detector. Geometry requires that $M \geq 1$, and when $M > 1$, the projected image is larger than the object. $M > 1$ can provide increased resolution if the detector is the limiting factor. Likewise, $h(x, y)$ and $H(u, v)$ refer to the source or focal spot distributions where the source is magnified by a factor of $M - 1$ onto the plane of the detector. The negative signs in the arguments of h and H indicate that the image of the source distribution is spatially reversed. When $M > 2$, the focal spot image is magnified potentially reducing detail in the final image. The trade-off between object magnification and focal spot magnification at the detector plane will be investigated later in this chapter after we discuss the effects of detector unsharpness. *Recall that magnification or enlargement in the spatial domain is minification or contraction in the spatial frequency domain.*

6.2.2 Motion Unsharpness

Ideally, x-ray imaging assumes that the patient, radiation source, and detector system are all stationary with respect to one another during the radiographic exposure. However, this is not true except in extreme cases, that is, postmortem. When one or more of these components move during the exposure, the image is blurred due to "motion unsharpness." Patient motion is usually the main cause of motion unsharpness since movement of both the detector and the source can be controlled. The motion of the patient can be complex and uncontrollable, for example, the motion due to cardiac motion or peristaltic motion in the abdomen. Motion due to breathing can be managed by breath-holding during the x-ray exposure. While it is difficult to model patient motion exactly, a simple 1-D model can be used to demonstrate this effect.

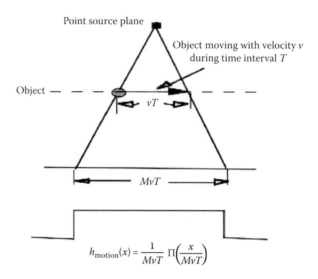

$$h_{\text{motion}}(x) = \frac{1}{MvT} \, \Pi\!\left(\frac{x}{MvT}\right)$$

FIGURE 6.6 Motion unsharpness is modeled as the loss of detail due to motion of a pinhole object during the radiograph exposure. For uniform velocity (v), the motion spread function has a rectangular shape.

Consider a point object (pinhole) moving uniformly in the x-direction at a constant velocity v during the exposure time T. The displacement of the projected image in the plane of the detector, due to motion of the object, is MvT, where M is object magnification. Under these assumptions, the pinhole image will take the form of a rectangle (Figure 6.6):

$$h_{\text{motion}}(x) = \frac{1}{MvT} \, \text{II}\!\left(\frac{x}{MvT}\right). \tag{6.15}$$

The term $1/(MvT)$ ensures that the motion spread function $[h_{\text{motion}}(x)]$ has unit area. Also, this ensures that it approaches a delta function in the limit where either v or T approach zero (little motion or extremely short exposure time). This motion spread function emphasizes that an effective and obvious method of limiting the effect of motion unsharpness is to use very short exposure times. For example, chest radiographs are taken with exposure times of 50 ms or less, to minimize the effect of cardiac motion and provide a moderately sharp image of cardiac borders.

6.2.3 Film-Screen Unsharpness

Unsharpness is also caused by light diffusion within the intensifying screen. When x-rays are absorbed at depth in the intensifying screen, the light diffuses, contributing to unsharpness. A thicker screen has more light diffusion and therefore will contribute more to unsharpness. The resolution varies from 6–9 lp/mm for images made with fast (i.e., thick) screens to 10–15 lp/mm for images obtained with detail screens under laboratory conditions. The resolution is usually much poorer (2–4 lp/mm) for fluoroscopic images obtained using an image intensifier. Light diffusion, and therefore screen unsharpness, can be limited in detail screens by tinting the phosphor layer of the intensifying screen. Do you know why?

An MTF of 10% is often used to represent limiting resolution. Unsharpness contributed by film, in most clinical examinations, is negligible in comparison with either geometric or screen unsharpness. For example, the MTF of film is ~50% at a spatial frequency of 20 lp/mm, while that of some screens is ~50% at about 1/10th this frequency (2 lp/mm), such that the loss in resolution is principally due to the screen at frequencies above 1–2 lp/mm. Traditional radiological practice uses screens for many diagnostic studies, so screen (not film) unsharpness is the major contributor to film-screen unsharpness. *So film is responsible for the non-linear response to radiation exposure and the screen is the major contributor to blurring.*

Following the arguments presented by Macovski and referring to Figure 6.7, we can calculate the spread function due to screens as a function of screen thickness. Assume that an x-ray photon enters the film-screen normal to its surface and is absorbed at a depth x in the intensifying screen. We will establish a coordinate system with an origin placed at the site of interaction. Let r be the distance within the plane of the film emulsion from the x-ray photon's path, and let d be the thickness of the intensifying screen.

Film exposure is proportional to the integrated fluence, the number of light photons per unit area reaching the emulsion. At $r = 0$, the position directly in line with the interaction site, the light fluence exposing the film depends on the depth of interaction x. Because light photons are assumed to be isotropically emitted from the interaction site, their fluence at $r = 0$ follows an inverse square law:

$$h(0,x) = \frac{k}{x^2} \qquad (6.16)$$

where k is a constant of proportionality that relates to the generation and propagation of light in the intensifying screen following the absorption of the x-ray photon. At a distance r from the origin, the light fluence falls off due to an inverse square law and also is modulated by a cosine term due to oblique angle that the light photons strike the film emulsion. Therefore,

$$h(r,x) = k\left(\frac{1}{r^2 + x^2}\right)\cos(\theta) = k\frac{x}{\left(r^2 + x^2\right)^{3/2}} \qquad (6.17)$$

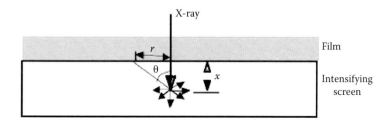

FIGURE 6.7   An x-ray photon is absorbed at a depth x in an intensifying screen. Light emitted by the phosphor at angle θ interacts with the film at a distance r from the path of the x-ray photon.

We assume that this system is linear and space-invariant so that the Fourier transform of Equation 6.17 gives the frequency response of the system. We can determine the spatial frequency behavior as radial frequency using the Hankel transform **H**, since $h(r, x)$ has circular symmetry (Appendix 6A).

Bracewell (p. 249) shows that

$$H\left[\frac{1}{a^2 + r^2}\right]^{\frac{3}{2}} = \frac{2\pi e^{-2\pi a\rho}}{a} \tag{6.18}$$

which can be applied to Equation 6.17 yielding

$$H_0(\rho, x) = H\{h(r, x)\} = 2\pi \int_0^\infty \frac{kx}{\left(\sqrt{r^2 + x^2}\right)^3} J_0(2\pi\rho r) r \, dr = \frac{2\pi e^{-2\pi x\rho}}{x} \tag{6.19}$$

where J_0 is the Bessel function of the zeroth order.

To standardize the frequency response, we will normalize H to give unit response at zero spatial frequency ($\rho = 0$) giving

$$H^*(\rho, x) = \frac{H_0(\rho, x)}{H_0(0, x)} = e^{-2\pi x\rho} \tag{6.20}$$

The function $H^*(\rho, x)$ gives the average frequency response per photon interacting at a depth x in the intensifying screen. When a beam of photons interacts with the intensifying screen, we must calculate the response due to all photons interacting with the screen, along the same vertical path, regardless of depth. If we assume that $p_d(x)$ is the fraction of all incident photons interacting per unit thickness at a depth x in the intensifying screen (i.e., the probability density function), then the average transfer function for this beam of interacting x-rays is

$$H(\rho) = \int_0^\infty H^*(\rho, x) p_d(x) \, dx = \int_0^\infty e^{-2\pi\rho x} p_d(x) \, dx \tag{6.21}$$

To determine $p_d(x)$, we observe that the fractional number of photons surviving after a distance x of an infinitely thick intensifying screen is $e^{-\mu x}$ so that the fraction of photons removed up to a distance x is

$$P(x) = 1 - e^{-\mu x} \tag{6.22}$$

$P(x)$ can be considered as the cumulative distribution function running from zero a $x = 0$ to 1 at $x = \infty$. The probability density function is the derivative of this cumulative distribution function:

$$p_d(x) = \frac{d}{dx}[P(x)] = \mu e^{-\mu x} \tag{6.23}$$

However, Equation 6.23 applies only to the unrealistic case of an infinitely thick intensifying screen. For an intensifying screen of finite thickness d, the fraction of photons removed up to a distance $x = d$ is

$$P'(x) = \frac{\text{Probability of photon interaction up to distance } x}{\text{Probability of photon interaction in screen of thickness } d} = \frac{1 - e^{-\mu x}}{1 - e^{-\mu d}} \tag{6.24}$$

where the aforementioned expression has been normalized using the number of photons absorbed in the screen so that $P'_d(x)$ represents the fraction of absorbed photons up to the distance x. This scaling normalizes the cumulative distribution function $P'(x)$ such that it ranges from zero at $x = 0$ to unity at $x = d$ as desired. The probability density function $p_d(x)$ for an intensifying screen of finite thickness d therefore is

$$P_d(x) = \frac{d}{dx}[P'(x)] = \frac{\mu e^{-\mu x}}{1 - e^{-\mu d}} \tag{6.25}$$

Based on Equation 6.21 the frequency response is as follows:

$$H(\rho) = \int_0^\infty H^*(\rho, x) p_d(x) dx = \frac{\mu}{1 - e^{-\mu d}} \int_0^d e^{-2\pi\rho x} e^{-\mu x} dx \tag{6.26}$$

$$H(\rho) = \frac{\mu}{(2\pi\rho + \mu)(1 - e^{-\mu d})} \left[1 - e^{-d(2\pi\rho + \mu)} \right] \tag{6.27}$$

Here, due to the assumed symmetry of the light production, the system transfer function H is also the system MTF. Note that as the radial frequency ρ approaches zero, then $H(\rho)$ approaches unity as required by the normalization of the psf.

For high spatial frequencies (where ρ is large), we know that

$$e^{-d(2\pi\rho + \mu)} \approx 0 \tag{6.28}$$

and that $2\pi\rho \gg \mu$, so we have at high frequency that

$$H_{hf}(\rho) \approx \frac{\mu}{(2\pi\rho)(1-e^{-\mu d})} \tag{6.29}$$

For the best spatial resolution, we want the magnitude of $H_{hf}(\rho)$ to be as large (i.e., close to 1) as possible at higher frequencies. In this regard, Equation 6.29 shows that there is a trade-off between screen thickness and spatial resolution. As the screen thickness d is increased, the efficiency of the screen increases reducing radiation dose to the patient. Alternatively, decreasing the screen thickness d increases the high-frequency response, increasing spatial resolution while requiring increased patient radiation dose. *It is important to note that phosphor materials with the higher absorption efficiency (i.e., larger values of μ) can have both a lower patient radiation dose and a higher high-frequency response.* Practically, even though the most efficient phosphors are used in modern intensifying screens, manufacturers offer intensifying screens having a range of phosphor thickness so that the practitioner can select the screen that offers sufficient spatial resolution while keeping patient exposure down.

6.2.4 Digital Image Resolution

We have represented an image $g(x, y)$ as the convolution between the object transmission $f(x, y)$ and the psf $h(x, y)$ of the imaging system:

$$g(x,y) = f\left(\frac{x}{M}, \frac{y}{M}\right) \otimes \otimes h\left(\frac{-x}{M-1}, \frac{-y}{M-1}\right) \tag{6.30}$$

where both the object and the psf are corrected for magnification M when projected onto the plane of the image detector. (*Note:* The psf is assumed to be normalized here so that we do not have to keep track of other scale factors.) The object term may contain blurring due to patient motion or other contributors to motion unsharpness, although these factors will be ignored in this section. Our goal here is to represent the digital image mathematically in terms of the analog image $g(x, y)$ and to evaluate the frequency characteristics of the digitized signal.

A two-step process can represent digitization, where we first *sample* the analog image and then *pixellate* the sampled image data. When we sample the image, we select regularly spaced values from the analog values that we assign to rectangular pixels during the pixellation process. In one dimension, let $g(x)$ represent the analog image from which we wish to generate the digital image with a pixel spacing of a. Sampling is done by multiplication of the image $g(x)$ by the comb function $III(x/a)$, and pixellation is done by convolving the sampled image with a rect function $II(x/a)$:

$$g_{dig}(x) = [g(x)III(x/a)] \otimes II(x/a) \tag{6.31}$$

where width a of the rect function is selected to exactly match the pixel/sample spacing (Figure 6.8).

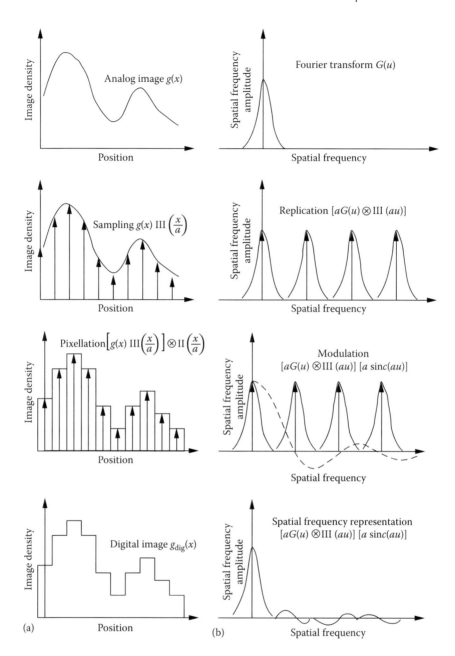

FIGURE 6.8 The process of image digitization begins with *sampling* of an analog image followed by *pixellation*. Each step in this process is illustrated for both spatial (a) and spatial frequency domains (b).

Intuitively, we suffer a loss of spatial resolution if pixel size and spacing are large compared to details we wish to preserve in the image. The sampling process can also introduce aliasing (i.e., a false frequency signal) if the sampling frequency does not satisfy Shannon's theorem (i.e., the sampling frequency must be twice the highest spatial frequency to be sampled).

Several effects of the digitization process are best seen using the frequency domain representation of the digitization, as follows:

$$G_{\mathrm{dig}}(u) = a^2 \left[G(u) \otimes \mathrm{III}(au) \right] \sin c(au) \tag{6.32}$$

In this equation, we see that in the frequency domain, digitization begins by convolving the Fourier transform $G(u)$ of the image function with the comb function $\mathrm{III}(au)$. Since the comb function is the sum of an infinite number of regularly spaced delta functions, this convolution replicates $G(u)$ at equally spaced intervals $\Delta u = 1/a$ on the spatial frequency axis. The replicates of $G(u)$ are further modified by multiplication by the sinc function. This multiplication reduces the amplitude of the higher spatial frequency replicates, since the magnitude of $\sin c(au)$ decreases with increasing frequency.

Note that the $\sin c(au)$ function has zeros that exactly coincide with the peaks in the replicates. However, this does not remove the higher frequency range in the tails of the distributions. It is just these high frequencies that lead to the box-like representation of original smooth analog signal. Compare the top left graph with the bottom left graph in Figure 6.8.

The replication term $[G(u) \otimes \mathrm{III}(au)]$ suggests the possibility of aliasing in the sampled function $[g(x)\, \mathrm{III}(x/a)]$. More specifically, if adjacent replicates of $G(u)$ overlap in the frequency domain, then aliasing occurs (Figure 6.9, lower right). If the highest spatial frequency in the analog image is u_{obj} [lp/distance] and the sampling frequency is

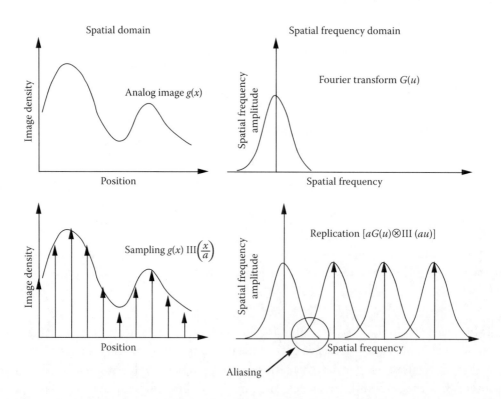

FIGURE 6.9 Aliasing is seen as the overlapping of replicates when the sample spacing is too large.

u_{samp} [samples/distance], then if $u_{\text{samp}} < 2u_{\text{obj}}$, then $G(u)$ is aliased. This shows up as the overlapping of the replicates in the frequency interval from $u_{\text{samp}} - u_{\text{obj}}$ up to the maximum frequency of $u_{\text{obj}}/2$. Figure 6.9 illustrates the origin of Shannon's sampling theorem from Chapter 3 that the signal must be sampled at a frequency twice the highest frequency in the signal, that is, to avoid aliasing or overlapping of replicates. This figure also illustrates that the frequency of the aliased signal equals the sampling frequency (u_{samp}) minus the presampled frequency (u_{obj}).

Aliasing can be prevented by smoothing (or "band-limiting") the analog signal $g(x)$ before sampling if the maximum frequency in the smoothed image is reduced to less than 1/2 the sampling frequency.

Mathematically, if smoothing is performed by convolving the analog image $g(x)$ with a low-pass filter kernel $b_{\text{lp}}(x)$, the digitized image can be expressed as

$$g_{\text{dig}}(x) = \left\{ \left[g(x) \otimes b_{\text{lp}}(x) \right] \text{III}(x/a) \right\} \otimes \text{II}(x/a) \qquad (6.33)$$

which in the spatial frequency domain is equivalent to

$$G_{\text{dig}}(u) = a^2 \left\{ \left[G(u) B_{\text{lp}}(u) \right] \otimes \text{III}(au) \right\} \text{sinc}(au) \qquad (6.34)$$

where $B_{\text{lp}}(u)$ is the Fourier transform of the smoothing or blurring function $b_{\text{lp}}(x)$. The blurring process has two important advantages. First, it can improve the signal-to-noise ratio of the digitized image by removing high spatial frequency components that often are dominated by noise. Second, blurring can reduce or eliminate aliasing. Of course, the function $b_{\text{lp}}(x)$ must be chosen carefully to reduce aliasing and improve the signal-to-noise ratio of the digitized image, without unacceptable loss of spatial resolution.

6.3 TRADE-OFFS BETWEEN GEOMETRIC UNSHARPNESS AND DETECTOR RESOLUTION

There is a fundamental trade-off between the increase in object detail that can be achieved due to object magnification and the decrease in object detail due to increased geometric unsharpness (source magnification). Object detail increases with increasing magnification if system resolution is limited due to poor spatial resolution of the detector. On the other hand, object detail decreases due to the magnified focal spot, so we seek a trade-off value for M that maximizes the detail obtainable.

If an image is recorded with object magnification M, and if the focal spot is modeled as a rectangle of width a, the width of the focal spot when projected onto the detector plane is $(M-1)a$. This width is equals $\dfrac{(M-1)a}{M}$ when back-projected to the object plane. We will characterize geometric unsharpness in the object plane using a cutoff frequency approach. The cutoff frequency is defined as that frequency where $H(u)$ first goes to zero. Since $h(x)$ is a rectangle function, $H(u)$ will be a sinc function, and we know that 1/width of a rectangle function will be at the first zero (cutoff frequency) of the sinc function. Therefore, the source cutoff frequency $u_s = 1/\text{width}$ and in the object plane is

$$u_s = \frac{M}{(M-1)a} \tag{6.35}$$

Similarly, detector resolution modeled as a rectangle of width b in the plane of the detector corresponds to a width of b/M in the object plane, giving a detector cutoff frequency in the object plane of

$$u_d = \frac{M}{b} \tag{6.36}$$

We can see the trade-off between u_d and u_s with increasing object magnification by graphing them as a function of object magnification (Figure 6.10).

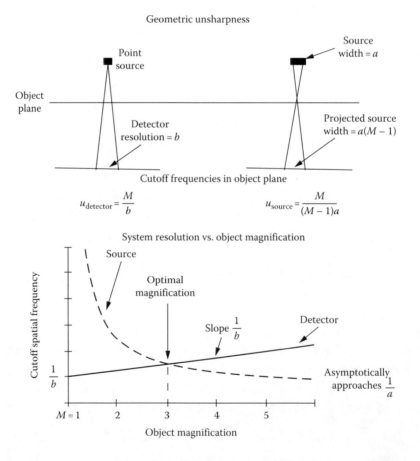

FIGURE 6.10 The source cutoff frequency (u_s) decreases and the detector cutoff frequency (u_d) increases with increasing magnification M. The component with the lowest cutoff frequency determines the overall cutoff frequency.

The curve labeled "source" shows how increasing magnification leads to resolution loss (lower cutoff frequency). In the limit of infinite magnification, the unsharpness due to the source asymptotically approaches the value

$$u_s = \frac{1}{a}\left(\lim M \to +\infty\right) \tag{6.37}$$

The curve labeled "detector" shows how increasing magnification leads to resolution gain (higher cutoff spatial frequency) with increasing magnification. At unit magnification, the cutoff frequency for the detector is $u_d = 1/b$, and for larger object magnifications, the cutoff frequency increases with a slope of $1/b$. At any magnification (M), the lower "detector" or "source" cutoff frequency determines the system cutoff frequency. Since the source cutoff frequency is monotonically decreasing with magnification and the object cutoff frequency is monotonically increasing, the maximum or optimal cutoff frequency occurs where their response curves cross (are equal). This happens when $b = (M-1)a$. This is the magnification at which the imaging system should be operated since it yields the highest cutoff frequency. *At this magnification, the magnified source width is equal to the detector resolution width.*

The optimal combination of digitization, detector, and geometrical components of resolution (or unsharpness) is explored in Example 6.1.

Example 6.1: Cutoff Frequency Analysis

Assume we are imaging a patient with an image intensifier system. The image intensifier has a resolution width of 0.2 mm and uses an x-ray tube with a focal spot width of 1.0 mm (both modeled as rectangle functions). The diameter of the image intensifier input phosphor is 15 cm, and the video output is delivered to an image processor that can produce digital images in either a 512 × 512 format or a 1024 × 1024 format. (For this problem, we will ignore the spatial resolution characteristics of the television camera, although generally this is an important consideration.)

(a) As a function of object magnification, graph the cutoff frequency for geometric unsharpness, detector unsharpness by the image intensifier, and unsharpness due to both the 512 × 512 image matrix and the 1024 × 1024 image matrix.

(b) From the graph generated in (a), determine the highest possible cutoff spatial frequency first for a 512 × 512 image and then for a 1024 × 1024 image. In each case, specify the object magnification at which this optimal system response occurs.

Solution

(a) Let M be the object magnification. The x-ray tube has a focal spot width of 1.0 mm. Therefore, its object cutoff spatial frequency is given by

$$u_s = \frac{M}{(M-1)(1.0\ \text{mm})} = \frac{M}{(M-1)}\ \text{mm}^{-1} \tag{6.38}$$

The image intensifier has a resolution width of 0.2 mm and therefore has an object cutoff spatial frequency of

$$u_d = \frac{M}{0.2 \text{ mm}} = 5.0M \text{ mm}^{-1} \tag{6.39}$$

The 512×512 matrix has a pixel width of

$$a_{512} = \frac{150}{512} \text{ mm} = 0.293 \text{ mm} \tag{6.40}$$

This is modeled as a rectangle of width 0.293 mm/M in the object plane with a cutoff spatial frequency of M/0.293 mm = $3.41M$ mm^{-1}. However, a more serious cutoff occurs at half this frequency due to sampling, that is, the Nyquist frequency limit, so the actual cutoff frequency is

$$u_{512} = \frac{256 \text{ cycles}}{150 \text{ mm}/M} = 1.71M \text{ mm}^{-1} \tag{6.41}$$

Finally, the 1024×1024 matrix has a pixel width of

$$a_{1024} = \frac{150}{1024} \text{ mm} = 0.146 \text{ mm} \tag{6.42}$$

with a corresponding Nyquist cutoff spatial frequency of

$$u_{1024} = \frac{512 \text{ cycles}}{150 \text{ mm}/M} = 3.41M \text{ mm}^{-1} \tag{6.43}$$

The graph of cutoff frequencies for the focal spot, image intensifier, 512×512 image matrix, and 1024×1024 image matrix is shown in Figure 6.11.

(b) When a 512×512 matrix is used, the detector resolution is limited by the matrix and focal spot size (geometric unsharpness) rather than by the image intensifier. The optimal system response is achieved when the cutoff spatial frequency from the 512×512 image matrix equals that from the x-ray tube focal spot. This condition is determined by setting $u_{512} = u_d$ and solving for M

$$M = 1.586 \tag{6.44}$$

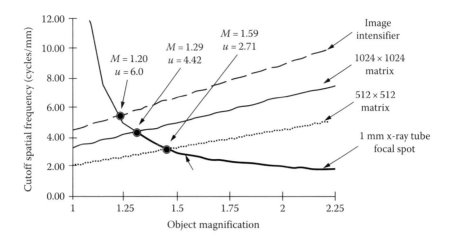

FIGURE 6.11   In Example 6.1, at low object magnifications (M), the spatial resolution of the system is limited by digitization. The 1024 × 1024 matrix is better than the 512 × 512 matrix. At higher object magnification (M), the spatial resolution is limited by blurring due to the 1 mm focal spot.

with a cutoff spatial frequency of

$$u_s = \frac{M}{M-1}\ \text{mm}^{-1} = \frac{1.586}{0.586} = 2.71\ \frac{\text{lp}}{\text{mm}} \tag{6.45}$$

Even when a 1024 × 1024 matrix is used, the spatial resolution is still limited by the image matrix and geometric unsharpness. The optimal system response is similarly achieved when the cutoff frequency from the 1024 × 1024 image matrix equals that due to the x-ray tube focal spot (geometrical component). This condition is met when

$$M = 1.292 \tag{6.46}$$

giving a cutoff spatial frequency of

$$u_s = \frac{M}{M-1}\ \text{mm}^{-1} = \frac{1.292}{0.292} = 4.42\ \frac{\text{lp}}{\text{mm}} \tag{6.47}$$

However, the system cutoff frequency increases dramatically *and is achieved with less magnification*.

6A APPENDIX: THE HANKEL TRANSFORM AND THE BESSEL FUNCTION

It often is convenient to express equations and formulae in terms of polar coordinates (r, θ) rather than Cartesian coordinates (x, y). If we know the Cartesian coordinates of a point (x, y), we can obtain its polar coordinates through well-known transformations

$$r^2 = x^2 + y^2 \tag{6A.1}$$

and

$$\theta = \arctan\left(\frac{y}{x}\right). \tag{6A.2}$$

Certain two-dimensional functions have circular symmetry. That is, if $f(x, y)$ is a function of the two independent Cartesian coordinates (x, y), when expressed in terms of polar coordinates, the function f can be specified entirely as a function of the radial coordinate r so that

$$f(r) = f(x, y) \tag{6A.3}$$

An example of such a function would be

$$f(x, y) = \exp[-\pi(x^2 + y^2)] \tag{6A.4}$$

which in polar coordinates is given by

$$f(r, \theta) = \exp[-\pi r^2] \tag{6A.5}$$

Obviously, $f(r, \theta)$ is circularly symmetric because it is independent of the angular variable θ.

If we can express a two-dimensional function f in terms of the single polar coordinate r (i.e., if f is circularly symmetric), then we can transform the Fourier transform from Cartesian to polar coordinates. In this case, the Fourier transform becomes the Hankel transform. If ρ is the spatial frequency variable corresponding to the spatial variable r, then the Hankel transform is defined by

$$F(\rho) = 2\pi \int_0^\infty f(r) J_0(2\pi\rho r) r\,dr \tag{6A.6}$$

while the inverse Hankel transform is defined as

$$f(r) = 2\pi \int_0^\infty F(\rho) J_0(2\pi\rho r) \rho\,d\rho \tag{6A.7}$$

where ρ is the conjugate spatial frequency variable for the spatial variable r (Equation 6A.1) and

$$\rho^2 = u^2 + v^2 \tag{6A.8}$$

To derive Equation 6A.6, we change to polar coordinates in the definition of the Fourier transform and integrate over the angular variable. That is, if

$$x = r\cos\theta \quad \text{and} \quad y = r\sin\theta \tag{6A.9}$$

are expressed in terms of the polar coordinates (r, θ), while the conjugate spatial-frequency variables

$$u = \rho \cos \phi \quad \text{and} \quad v = \rho \sin \phi \tag{6A.10}$$

are expressed in terms of the polar coordinates (ρ, ϕ), then

$$F(u,v) = \int_{-\infty}^{+\infty} \int_{-\infty}^{+\infty} f(x,y) e^{-2\pi i(ux+vy)} dx dy \tag{6A.11}$$

$$F(u,v) = \int_{-\infty}^{+\infty} \int_{-\infty}^{+\infty} f(x,y) e^{-2\pi i r\rho(\cos\theta\cos\vartheta + \sin\theta\sin\vartheta)} dx dy \tag{6A.12}$$

$$F(u,v) = \int_{-\infty}^{+\infty} \int_{-\infty}^{2\pi} f(r) e^{-2\pi i r\rho\cos(\theta-\vartheta)} d\theta r \, dr \tag{6A.13}$$

$$F(u,v) = \int_{0}^{+\infty} \int_{0}^{2\pi} f(r) e^{-2\pi i r\rho\cos\theta} d\theta r \, dr \tag{6A.14}$$

And using the relationship defining the zero-order Bessel function of the first kind $J_0(z)$ where

$$J_0(z) = \frac{1}{2\pi} \int_{0}^{2\pi} e^{iz\cos\theta} d\theta \tag{6A.15}$$

the definition of the Fourier transform in polar coordinates for circularly symmetric functions becomes the Hankel transform $H(\rho)$ where

$$H(\rho) = 2\pi \int_{0}^{+\infty} f(r) J_0(2\pi\rho r) r \, dr \tag{6A.16}$$

HOMEWORK PROBLEMS

P6.1 The value of the MTF of a system at 2 lp/mm is 0.25. Assuming three serial components contribute to system blurring and that two are identical with an MTF value of 0.7 at 2 lp/mm, calculate the MTF value of the third component at that frequency.

P6.2 During an experiment to measure the spatial frequency response of an imaging system, we obtain an esf, given by the normalized measurements in the following graph. Assume that the image unsharpness is due to penumbral effects from the source and that detector unsharpness is negligible, and that the image is recorded with an object magnification $M = 2$.

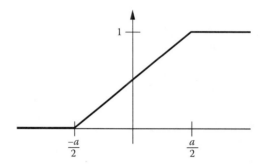

(a) From the esf shown earlier, show that the line spread function lsf(x) for the system is

$$\text{lsf}(x) = \frac{1}{a} \Pi\left(\frac{x}{a}\right)$$

(b) Assume that we have a source distribution of

$$t_s(x) = \frac{1}{a} \Pi\left(\frac{x}{a}\right)$$

and a new detector, which has a detector psf described by a rectangular function

$$t_d(x) = \frac{1}{b} \Pi\left(\frac{x}{b}\right)$$

When the effects of both geometric unsharpness and detector unsharpness are included, show that the MTF of the system is

$$\text{MTF}_{\text{system}}(u) = |\sin c(au)\sin c(bu)|$$

Moreover, if $a = b$, show that the psf of the system is given by the equation

$$\text{psf}(x) = \frac{1}{a} \Lambda\left(\frac{x}{a}\right)$$

(c) If $a \neq b$, sketch the psf of the system and briefly justify your answer conceptually and mathematically.

P6.3 A resolution pattern consists of equal width thin lead bars separated by equal widths of a radiolucent material (air, plastic, or thin aluminum). Assume that the spacing between adjacent lead bars equals the distance a and that the width of the interspace material and of the lead bars is equal.

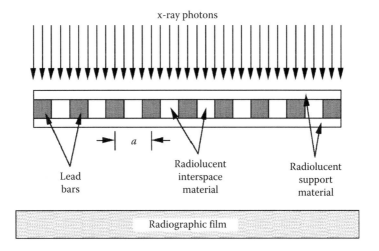

Assume that we fix the distance between the x-ray source and the sheet of film at a distance L. When the pattern is placed directly on the film, we can resolve the individual bars. However, bringing the bar pattern closer to the focal spot increases geometric unsharpness of the bar pattern until its appearance on the radiograph is completely suppressed. The position at which this suppression occurs is found to be the distance h as measured between the resolution pattern and the source.

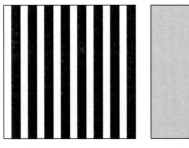

Appearance of the resolution phantom when it is placed directly on the film.

Appearance of the resolution phantom when it is located at a distance h from the x-ray source.

(a) Show that the projected spacing s of the bars in the resolution pattern, in terms of a, L, and h, is

$$s = \frac{aL}{h}$$

(b) Assuming a rectangular focal spot distribution. Show that the width of the focal spot can be described by the function

$$h(x) = \frac{1}{b} \text{II}\left(\frac{x}{b}\right)$$

where

$$b = \frac{aL}{L - h}$$

(c) Assuming that detector unsharpness is negligible, show that the MTF of the system is

$$\text{MTF}(u) = \left| \text{sinc}\left[\left(\frac{aL}{h}\right)u\right] \right|$$

(d) From the MTF determined in (c), describe why the bar pattern disappears in the image.

(e) Draw a diagram showing the path of the x-rays from a rectangular source as they irradiate the bar pattern. Use this diagram to present an intuitive explanation how geometric unsharpness suppresses the image of the bar pattern in the radiograph.

P6.4 A hypothetical defective image intensifier produces 3 parallel thin lines on the output screen when a thin line phantom is imaged. If the lines are separated by distance a, show that the MTF is given by

$$\text{MTF}(u) = \frac{1}{3} + \frac{2}{3}\cos(2\pi u a)$$

where u is spatial frequency in units of lp/mm.

P6.5 If a rectangular focal spot has a width equal to a and the detector has a resolving width of b, show that the optimum system resolution is obtained at a magnification of

$$M = \frac{b}{a} + 1$$

Discuss conceptually why this result is reasonable.

P6.6 We are imaging a patient using a scanning detector array system with a linear array of discrete scintillation detectors. The width of each detector element in the array is 5 mm. The array is oriented vertically and is moved from left to right across the patient to create the image. The system has an x-ray tube in which the operator can select one of two focal spot sizes. The large focal spot has a width of 1.6 mm. The small focal spot has a width of 0.6 mm.

(a) As a function of object magnification, graph the cutoff spatial frequencies for detector unsharpness and for geometric unsharpness contributed by both the large and small focal spots (i.e., 3 separate curves: one for the detector, one for the small focal spot, and one for the large focal spot).

(b) Use the graph in (a) to decide the "optimal" object magnification (i.e., which reproduces the highest spatial frequencies in the patient) first for the large focal spot and then for the small focal spot. In each case, what is the highest cutoff spatial frequency and the optimal magnification for the system?

(c) Assume that the detector array is located 2 m from the x-ray source. To achieve optimal spatial resolution for structures in the patient, how far should the patient be placed in front of the detector array first for the large focal spot and then for the small focal spot?

(d) As we mentioned at the beginning of this problem, the width of each detector element is 5 mm. Assume we wish to scan across a 35 cm region of the patient in 2 s. How fast should the detector data be sampled to produce pixels that represent square regions on the patient?

P6.7 You are digitizing an image described in one dimension by $f(x)$. The sampling distance is chosen to equal the pixel width. Assume that you have a square image covering 25 cm on each side and that you are digitizing the image with a 512×512 matrix. To prevent aliasing, your instructor tells you to convolve the image with a sinc function of width with zero crossing points at the points

$$x_n = nb \quad \text{where } n = \pm 1, \pm 2, \ldots$$

where b is a constant distance.

(a) First, from the information presented earlier, derive the equation of the sin c function in terms of b that you will use to convolve the image prior to sampling and pixellation.

(b) To analyze the effects of this operation, present a diagram analogous to that presented in the notes showing the effect of low-pass filtering, sampling, and pixellation. Diagram the effects in both the spatial and the spatial frequency domains. Briefly explain each step of the diagram.

(c) Find the minimum value of b that will prevent aliasing but will minimize the loss of spatial frequency information in the signal.

(d) Describe the effect on the digital image if, before sampling, you convolve the image with (i) a Gaussian or (ii) a square function instead of a sin c function.

P6.8 In a cineradiographic study of the heart, the maximum speed of motion of the coronary arteries is 10 cm/s. If the heart is imaged with a magnification of 1.3 and an exposure time of 4 ms, calculate the physical width of a rectangular focal spot that will give an MTF at the image intensifier similar to that due to the blurring due to heart motion. (Assume the image is captured at the onset of systole and during which the myocardium moves uniformly at its maximum velocity.)

The primary question we face is how do we select the exposure time to minimize image unsharpness. If we use a small focal spot to decrease geometric unsharpness, we must increase the exposure time that increases motion unsharpness. Alternatively, if we minimize motion unsharpness by shortening the exposure time, this forces us to use a larger focal spot that increases geometric unsharpness. Faced with this trade-off, we must ask how we can select the focal spot width to minimize geometric unsharpness but which also allows us to minimize the exposure time to limit motion unsharpness.

(a) Assume that we need an amount of photon energy equal to E_s to obtain a satisfactory image. Moreover, we operate the x-ray tube at its maximum power density of P_{max}. If the focal spot is square in shape with width L, show that the shortest possible exposure time is

$$T = \frac{E_s}{P_{max} L^2}$$

(b) Assume that the source, described by the function $s(x, y)$, emits x-rays, which pass through a stationary object with an transmission function $t(x, y)$. If the object magnification is M, the image $i(x, y)$ is given by the convolution of s with t, with factors to correct for magnification of both the object and the source.

$$i(x,y) = t\left(\frac{x}{M}, \frac{y}{M}\right) \otimes \otimes s\left(\frac{x}{M-1}, \frac{y}{M-1}\right)$$

First, assume that the object moves with uniform velocity v during the exposure time T. Second, assume that the source is square in shape with width L. In this case, using the result from part (a), describe why the image $i(x, y)$ is described by the function

$$i(x,y) = t\left(\frac{x}{M}, \frac{y}{M}\right) \otimes \otimes \Pi\left(\frac{x}{(M-1)L}, \frac{y}{(M-1)L}\right) \otimes \left[\frac{P_{max}L^2}{MvE_s}\Pi\left(\frac{P_{max}L^2 x}{MvE_s}\right)\right]$$

(c) Use the result from part (b) to show that the width X of the system psf including the effects of motion unsharpness and focal spot blurring is approximated by

$$X = (M-1)L + \frac{MvE_s}{P_{max}L^2}$$

Show that the focal spot width L_{min} that minimizes the width of the system psf is

$$L_{min} = \left[\frac{2MvE_s}{P_{max}(M-1)} \right]^{\frac{1}{3}}$$

(d) Assume we need a source energy E_s of 600 J to generate an image at a power density P_{max} of 10^7 W/cm². At a magnification M of 1.5 and a velocity v of 10 cm/s, calculate the width of the focal spot and the corresponding exposure time that minimizes system unsharpness.

(e) If we have a constant potential x-ray generator operating at 100 kV, calculate the tube current that delivers the required x-ray dose to the patient to generate the image under the conditions you have derived in part (d).

P6.9 Here is a problem for fun that does not have anything to do with spatial resolution or MTF. One day I was trying to explain to my cousin and good friend Sachi how inefficient x-ray tubes are at generating x-rays, especially when used in a fan beam geometry. For some reason, Sachi did not find this conversation very exciting. However, because Sachi has just graduated from UCLA and is a little short on cash right now, she appreciated it a little more when I did a financial analysis about x-ray tube efficiency for her. Let us repeat this calculation for Sachi.

(a) Assume that an x-ray tube, having a tungsten anode, is operated at 100 kVp and 100 mA. If the tube is on for a total of 2 h/day for 6 days a week and 52 weeks per year, calculate how much energy is consumed by the x-ray system in 2 year.

(b) The fraction of this energy that is used to generate x-rays (instead of heat) is given by the formula

$$\text{Fractional efficiency} = 0.9 \times 10^{-9} \, VZ$$

where
V is the x-ray tube potential in volts
Z is the atomic number of x-ray tube target

Use your result from part (a) to calculate how much energy is liberated in the form of usable x-rays over 1 year.

(c) The x-rays from the anode are given off isotropically. Assume we collimate the beam to irradiate a detector array 5 mm wide and 43 cm long 2 m from the source. What fraction of the total x-ray beam is used to irradiate the detector array? Over the period of 1 year, how much x-ray energy is emitted in the direction of the detector array?

(d) Assuming that electricity from your local supplier costs 12¢ per kilowatt-hour, calculate the total cost of electricity consumed by the x-ray tube over the period of 1 year. Also calculate the "cost" of the photons delivered to the detector array in the form of usable x-rays. (Sachi will be appalled at the difference between these two amounts!)

REFERENCE

Bracewell, RN. *The Fourier Transform and Its Applications.* 2nd eds., McGraw-Hill, Inc., New York, 1986.

CHAPTER 7

Random Processes

S O FAR, WE HAVE dealt with the emission and attenuation of radiation as if they were entirely predictable processes, and unfortunately, they are not. For example, suppose the detection system records the number of photons from a radiation source striking a detector in a specified time interval. If this counting experiment is repeated multiple times, the number of counts recorded will vary. In fact, the number of counts would be quite random, and even if you knew exactly the number of counts recorded in one trial, you could not predict the number of counts in the next or in any subsequent one. To paraphrase Einstein, God apparently does play dice with the universe, which does not behave chaotically. However, the measurements of counts in this example are randomly distributed about some average value. If a large number of measurements are made, you can determine an average (or mean) number of counts. You can also estimate the standard deviation, which indicates the spread in the number of counts about the mean value. You can determine the "probability" that a certain range in the number of counts would be recorded for any time interval, but you cannot predict the exact count. Therefore, while you can determine statistical descriptors, such as the mean value and standard deviation, using a large number of measurements, you cannot predict individual measurements. This is the nature of measurements associated with any random process.

Like emission of radiation, attenuation of radiation is a random process; if you could deliver the same number of photons to an object, the number of photons transmitted through that object would vary randomly. Again, after making enough measurements, you would be able to estimate the mean and standard deviation in the number of photons transmitted, and determine the probability of a given range in the number of photons to be transmitted through the object, but cannot predict the exact number. This is true even if you know exactly how many photons were delivered to the object.

Because the basic processes of emission and attenuation of radiation are random, and because the random nature of these processes is inherent and physically unavoidable, we are compelled to study the nature and characteristics of these random processes. This is very important, since it focuses our attention on the uncertainty in images that we will associate with random noise. We previously defined random noise as the random variation in a signal (i.e., in an image). As we will discover in the next chapter, this random variation has several physical causes. Whether dealing with emission imaging with a radiotracer as the source, transmission imaging where the source is the focal spot of an x-ray tube, or the emitted RF signal for magnetic resonance imaging, we must understand features and limitations imposed by the random nature of these imaging modalities' radiations. For this, we must develop the tools to model and quantify random processes that relate to noise in images.

7.1 PROBABILITY DISTRIBUTIONS

Assume that we perform an experiment during which we derive an outcome x. Furthermore, assume we know that a specific random process determines this outcome. For example, our outcome may be number of photons that we measure from a radiation source in a fixed time interval. This is an example where possible outcomes are discrete; we may obtain 1 count or 100 counts but will never obtain partial (e.g., 3.25) counts in any given measurement. In a second example, our outcome might be the distance that a 60 keV x-ray photon travels through a column of water before it is Compton scattered. In this second example, the possible outcomes are continuous and may take on any value greater than zero. Whether we are dealing with discrete or continuous measurements, we would like to know the "probability" of a given outcome. Probabilities can be calculated using mathematical functions called probability distribution functions.

7.1.1 Discrete Probability Distributions

In the case of a discrete variable n (i.e., one where outcomes are integer values), the probability distribution of outcome n, denoted as $P(n)$, gives the probability that an experiment will produce an outcome equal to n. For example, when recording counts from a radioactive source using a fixed counting time, if $P(5) = 0.12$, we are stating that in 12% of the trials, we expect to measure 5 counts.

A characteristic feature of a discrete probability distribution is that the summed probability is normalized to be equal to 1 for all possible outcomes. Mathematically, this is given by the following equation:

$$\sum_{n=-\infty}^{+\infty} P(n) = 1. \tag{7.1}$$

We define the mean value of n as μ and variance of n as σ^2 for the random variable n, which is distributed according to the probability distribution $P(n)$ using the following equations:

$$\mu = \sum_{n=-\infty}^{+\infty} nP(n). \tag{7.2}$$

$$\sigma^2 = \sum_{n=-\infty}^{+\infty} (n-\mu)^2 P(n). \tag{7.3}$$

In these equations, it is easy to see that μ is just the average value of n weighted by $P(n)$. The variance (σ^2) is then the average value of $(n - \mu)^2$, the square of the distance of n from the mean. The standard deviation (σ) is defined as the square root of this variance. The use of Greek symbols for mean and variance apply when $P(n)$ covers all possible values of n, and these are called population mean and variance. When working with a limited sample (not covering all possible values of n), the mean is called sample mean ($\bar{n}$) and the variance is called sample variance (s^2), with non-Greek symbols to distinguish them from population parameters.

7.1.2 Continuous Probability Distributions

In the case of a continuous variable x, the definition of the probability distribution of x, denoted by $P(x)$, is defined slightly differently from that given for the discrete random variable. In particular, we define the probability distribution $P(x)$ such that $P(x)dx$ gives the probability for the interval from x to $x + dx$. For continuous random variables, we call $P(x)$ the probability "density" function, sometimes abbreviated pdf(x).

Continuous probability distributions, like the discrete probability distributions, have certain characteristics that are important. In particular, we know that the probability is equal to 1 when we consider all possible outcomes. Mathematically, this means that integration from $x = -\infty$ to $x = +\infty$ is unity:

$$\int_{x=-\infty}^{+\infty} P(x)dx = 1. \tag{7.4}$$

We define the mean value μ and variance σ^2 of the continuous random variable x distributed according to the probability density function (pdf) $P(x)$ as

$$\mu = \int_{x=-\infty}^{+\infty} xP(x)dx, \tag{7.5}$$

$$\sigma^2 = \int_{x=-\infty}^{+\infty} (x-\mu)^2 P(x)\,dx. \tag{7.6}$$

Again, the standard deviation is σ.

7.2 SPECIAL DISTRIBUTIONS

There are three probability distributions that we are particularly interested in because they are frequently encountered in medical imaging. They are (1) the binomial distribution, (2) the Poisson distribution, and (3) the Gaussian (or normal) distribution. These distributions are presented here without derivation, but derivations can be found in many statistics textbooks.

7.2.1 Binomial Distribution

The binomial distribution is a probability distribution of a discrete random variable that can have only two possible outcomes (called "success" or "failure"), where the probability of success in an individual trial is p, and therefore, the probability of failure is $1 - p$. The binomial distribution is signified by $P(m; n, p)$, which gives the probability of m successes out of n individual trials where p is the probability of success. The mathematical formula for the binomial distribution is

$$P(m; n, p) = \frac{n!}{m!(n-m)!} p^x (1-p)^{n-m}.$$ (7.7)

Note: The number of successes m is the discrete random variable, and n and p are distribution-specific parameters. An example where the binomial distribution can be used is calculating the probability that a certain number of heads will occur in repeated tosses of a fair coin (one for which heads and tails have equal probability). The probability of getting a head in one trial is $1/2$ ($p = 0.5$). Therefore, the probability of getting 2 heads ($m = 2$) in 6 trials ($n = 6$) is

$$P(2; 6, 1/2) = [6!/(2!\ 4!)]\ (1/2)^2 (1/2)^4 = 0.23.$$ (7.8)

When $p = 1/2$, then $(1 - p) = 1/2$; the last two terms in Equation 7.7 can be grouped as $(1/2)^n$. The probability of obtaining all tails ($m = 0$) or all heads ($m = 6$) in 6 trials is just $(1/2)^6$. *Note*: by definition $0! = 1$. Finally, the sum of Equation 7.7 over all possible outcomes is equal to one, since this is the probability of all possible outcomes.

7.2.2 Poisson Distribution

A second important probability distribution of a discrete random variable is the Poisson distribution where a single trial can produce zero through a positive number of successful results. Since this is a probability distribution of a discrete random variable, a fractional outcome is not permitted. If independent trials are conducted with a mean value $= \lambda$, then the probability that the outcome $= n$ in a single trial is given by the Poisson distribution function:

$$P(n; \lambda) = \frac{e^{-\lambda} \lambda^n}{n!}.$$ (7.9)

Unlike the binomial distribution, the Poisson distribution has a single parameter λ. Also, unlike n, which must be a positive integer, λ (also positive) can have a fractional value, since it is the average of many independent trials. For example, if we record a mean of 5.63 counts from many 1 s trials from a radioisotope counting experiment,

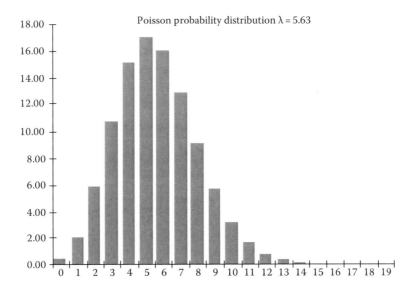

FIGURE 7.1   Poisson probability distribution. Vertical axis scaled to highlight relative values.

the number of counts for individual trials will be distributed according to a Poisson distribution with $\lambda = 5.63$ (Figure 7.1):

$$P(n) = \frac{e^{-5.63} 5.63^n}{n!}. \tag{7.10}$$

Therefore, the probability of obtaining 6 counts in a 1 s measurement is

$$P(6) = \frac{e^{-5.63} 5.63^6}{6!} = 15.87\%. \tag{7.11}$$

Inspection of Figure 7.1 shows that for this example, the most likely outcome for any 1 s measurement is $n = 5$. Also note that the distribution is skewed toward values greater than the mean. This asymmetry diminishes with increasing mean value.

The probability of obtaining a count of 5 or less $P(n \leq 5)$ would require a sum over the probabilities of obtaining counts ranging from 0 to 5. The probability of obtaining a count greater than 3 would therefore be $P(>3) = 1 - P(n \leq 3)$.

A fundamental property of the Poisson distribution is that the variance is equal to the mean, that is, $\sigma^2 = \lambda$. This is particularly important to radiation physicists, since processes such as nuclear disintegration and photon attenuation are Poisson-distributed processes. If we accurately determine the mean value of a random process that is Poisson-distributed, then the variance is equal to the mean and the standard deviation is just the square root of the mean value. For example, in radiation counting, a value of $n = 10,000$ counts is accepted as a reasonably accurate measurement of the mean value, because the estimated standard deviation for repeat counting would be $(10,000)^{1/2} = 100$, just 1% of the mean.

7.2.3 Gaussian Distribution

Unlike the binomial and Poisson distributions, the Gaussian distribution is a probability distribution of a continuous random variable x. The pdf for the Gaussian distribution is given by

$$\text{pdf}(x; \mu, \sigma) = \frac{1}{\sqrt{2\pi}\sigma} e^{-\frac{1}{2}\left(\frac{x-\mu}{\sigma}\right)^2}. \tag{7.12}$$

Unlike the other two distributions, the mean value $= \mu$ and the standard deviation $= \sigma$ are independent parameters in the Gaussian pdf. Also, unlike the other two distributions, Gaussian distributions are symmetric about the mean value.

Since the Gaussian distribution is a function of a continuous variable, probability must be calculated by integration over the range of interest. For example, the probability that a Gaussian-distributed random variable x falls within the limits $x = a$ to $x = b$ is

$$\text{Probability of } (a \le x \le b) = \int_a^b \text{pdf}(x; \mu, \sigma) dx. \tag{7.13}$$

Note that the continuous Gaussian pdf has units of probability/distance. An important characteristic of the Gaussian distribution is that 68.3% of outcomes fall within one ±1 standard deviation of the mean (i.e., in the interval from $\mu - \sigma$ to $\mu + \sigma$), 95.5% fall within ±2 standard deviations of the mean, and 99.7% fall within ±3 standard deviations of the mean.

7.2.3.1 Gaussian Distribution as a Limiting Case of the Poisson Distribution

Both Poisson and Gaussian distributions are important in x-ray imaging. This is because in the limit of large values of n, the Poisson distribution approaches a Gaussian distribution (Figure 7.2). For large values of n, substituting n for both μ for σ^2 in Equation 7.12 simplifies the resulting Gaussian formula:

$$\text{pdf}(x; n) = \frac{1}{\sqrt{2\pi n}} e^{-\frac{1}{2}\frac{(x-n)^2}{n}}. \tag{7.14}$$

The pdf in Equation 7.14 has both Poisson and Gaussian characteristics. Since we typically deal with large numbers of photons in radiographic imaging, Equation 7.14 can be used to describe the statistical nature of photon emission from x-ray and gamma-ray sources, as well as that of the attenuation of photons by matter. For example, we can apply properties

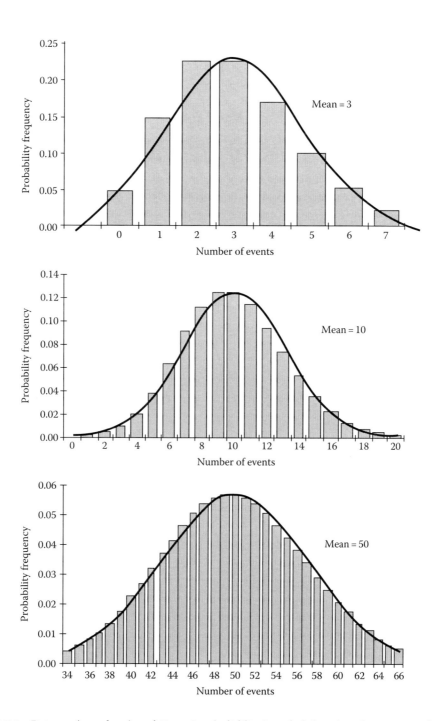

FIGURE 7.2   Poisson (gray bars) and Gaussian (solid line) probability distributions are shown for mean values of 3, 10, and 50. The Poisson probability distribution becomes more Gaussian-like as the mean value increases.

of a Gaussian distribution in that 68.3% of outcomes fall in the range $\mu \pm \mu^{1/2}$, 95.5% fall in the range $\mu \pm 2\mu^{1/2}$, and 99.7% fall in the range $\mu \pm 3\mu^{1/2}$. *Note*: In the case of a Poisson distribution that $\mu = n$, which is a dimensionless number, so there is no problem calculating the standard deviation as its square root.

The properties of a Poisson distribution are important in medical imaging, since we often characterize the level of noise as the standard deviation of a signal. If the signal is randomly distributed according to a Poisson distribution, the variance of the signal is equal to the mean. In counting experiments, if we obtain n counts in a measurement, colloquial expression of this concept is that "noise" or the standard deviation equals the square root of n. What is really meant by this expression is that (1) if one records a measurement of the number of photons recorded by an ideal detector for a fixed time period and measures n photons, and (2) if one assumes that the measurement reflects the mean number of counts recorded over a large number of measurements, and (3) if one assumes that the distribution of counts follows a Poisson distribution, and (4) if one has a large number of counts (>20) so that the distribution is approximately Gaussian, then the standard deviation is approximately equal to square root of n and 68.3% of future measurements should fall within +/− this standard deviation.

7.3 REVIEW OF BASIC STATISTICAL CONCEPTS

A review of several basic mathematical concepts associated with statistics will be helpful before introducing the valuable tools of Sections 7.4 and 7.5.

Concept 1: It is often necessary to determine how algebraic manipulation of a random variable affects the associated mean and variance. Let $y = ax + b$, where the random variable y is calculated from the random variable x. The expected or mean value of y where $E\{\}$ is the expectation operator is

$$E\{y\} = E\{ax + b\} = aE\{x\} + E\{b\}, \tag{P.1}$$

so $\mu_y = a\mu_x + b$. For this algebraic manipulation, μ_y is a times μ_x plus b. Note that if μ_x is zero, then μ_y is determined by b alone. An important case follows if $a = 1$ and $b = -\mu_x$ such that $y = x - \mu_x$. In this case, y will be zero mean, and importantly, any random variable can be adjusted to be zero mean using this approach, that is, by subtraction of its mean value.

The variance of $y = ax + b$ is calculated using the variance operator $V\{\}$ as follows:

$$V\{y\} = V\{ax + b\} = E\{[(ax + b) - (a\mu_x + b)]^2\} = E\{(ax - a\mu_x)^2 = a^2\, E\{(x - \mu_x)^2\} = a^2 V\{x\}. \tag{P.2}$$

So, $\sigma_y^2 = a^2 \sigma_x^2$.

Equation P.2 shows that $V\{y\}$ does not depend on b and scales by a^2. Importantly, if $a = 1/\sigma_x$, then σ_y will be unity. A combination of Equation P.1 (to set mean = 0) and Equation P.2 (to set variance = 1) is used to transform a Gaussian random variable to the normally distributed random variable z:

$$z = (x - \mu_x)/\sigma_x, \tag{P.3}$$

with mean = 0 and standard deviation = 1. Probabilities for the normally distributed z are found in all statistics texts and available in spreadsheet applications such as excel.

Concept 2: Another important concept deals with determining mean and variance when adding or subtracting random variables. In this case, where x and y are both random variables, the expected or mean value of $z = x \pm y$ is calculated as follows:

$$E\{z\} = E\{x \pm y\} = E\{x\} \pm E\{y\} \rightarrow \mu_z = \mu_x \pm \mu_y. \tag{P.4}$$

Equation P.4 states that the mean value resulting from adding or subtracting two random variables is the sum or difference in their mean values. Clearly, this result will be valid for adding or subtracting more than two random variables.

The calculation of variance of $z = x \pm y$ is a bit more complex as follows:

$$V\{z\} = V\{x\} + V\{y\} \pm 2\,\text{CoV}\{x, y\}. \tag{P.5}$$

The covariance term may or may not be zero but must be included in the formal math. The covariance between x and y is calculated as

$$\text{CoV}\{x, y\} = E\{(x - \mu_x)(y - \mu_y)\} = E\{xy\} - \mu_x\mu_y. \tag{P.6}$$

The variance in Equation P.5 differs depending on $\text{CoV}(x, y)$. If x and y are independent random variables, then $E\{xy\} = E\{x\}E\{y\} = \mu_x\mu_y$ such that $\text{CoV}\{x, y\} = 0$, and the net variance is just the sum of variances of x and y. Note that $V\{x\}$ and $V\{y\}$ in Equation P.5 are added whether z is calculated by adding or subtracting, unlike the covariance term, that is, *resulting variance will be larger when adding or subtracting random variables*. If $\text{CoV}\{x, y\}$ is positive (x and y rise above and below their mean values together), then $V\{z\}$ will be larger when adding but smaller when subtracting (do you know why?). If $\text{CoV}\{x, y\}$ is negative, then we observe the opposite effect for $V\{z\}$. The basic trend in Equation P.5 holds for more than two random variables, but covariance terms get a bit messy, so we will stick with just two. In many cases, we can assume that the random variables are either independent or uncorrelated such that the covariance term in Equation P.5 is zero and can be ignored.

Concept 3: It is important to understand what happens to the mean and variance when averaging a group of random variables. For example, the mean value of random variables (x_i) from a population x each with mean = μ and variance = σ^2 is

$$E\{x\} = 1/n\ E\{x_1 + x_2 + x_3 + \cdots + x_n\}, \tag{P.7}$$

$$E\{x\} = 1/n\ (E\{x_1\} + E\{x_2\} + E\{x_3\} + \cdots + E\{x_n\}), \tag{P.8}$$

$$E\{x\} = 1/n\ (n\mu), \tag{P.9}$$

$$E\{x\} = \mu. \tag{P.10}$$

So the mean value of from a set of random variables when averaging a large number of samples tends to the population mean. The variance of this set of random variables can be calculated using Equations P.2 and P.5 as follows where $E\{\}$ is shortened to $<>$ for clarity:

$$V\{<x>\} = V\{1/n <x_1 + x_2 + x_3 + \cdots + x_n>\} = (1/n)^2 \, V\{<x_1> + <x_2> + <x_3> + \cdots + <x_n>\}, \tag{P.11}$$

$$V\{<x>\} = (1/n)^2 \, (n\sigma^2), \tag{P.12}$$

$$V\{<x>\} = \sigma^2/n. \tag{P.13}$$

So though $E\{x\}$ tends toward the population mean when averaging, the variance is reduced by $1/n$ of the population variance. This reduction in variance of the mean with increasing n is used to establish group size when seeking to measure a significant difference between mean values of two groups, each with n random measurements.

7.4 CENTRAL LIMIT THEOREM

A very important theorem in statistics is the central limit theorem, which states that the distribution of "mean values" from any random distribution is approximately Gaussian, provided that the number of samples is large enough. For example, we can use the central limit theorem to generate a Gaussian distribution from the uniform distribution, which is definitely non-Gaussian. The random number generator found on most computers can provide samples from a uniform distribution. Mean values calculated as averages of "n" samples from these data will be Gaussian distributed.

The uniform distribution $u(x)$ defined for the continuous random variable x over the interval $[0, 1]$ is

$$u(x) = \begin{cases} 1 & \text{for } 0 \leq x \leq 1 \\ 0 & \text{eslewhere} \end{cases} \tag{7.15}$$

First, note that $u(x)$ is normalized so that the integral of $u(x)$ over all values of x is unity:

$$\int_{-\infty}^{+\infty} u(x)\mathrm{d}x = \int_0^1 \mathrm{d}x = 1. \tag{7.16}$$

Second, the mean μ of this uniform distribution is equal to

$$\mu = \int_{-\infty}^{+\infty} x \cdot u(x)\mathrm{d}x = \int_0^1 x\mathrm{d}x = \frac{1}{2} x^2 \Big|_0^1 = \frac{1}{2} \tag{7.17}$$

and the variance is

$$\sigma^2 = \int\limits_{-\infty}^{+\infty} (x-\mu)^2 u(x)\,dx = \int\limits_{0}^{+1} \left(x-\frac{1}{2}\right)^2 dx = \frac{1}{12}. \tag{7.18}$$

Using Equation P.1, we can generate a zero-mean distribution of random numbers r_i by subtracting the mean value, here ½, from the uniformly distributed x_i's. If we create 12 such zero-mean uniform distributions (perhaps with 100 samples each) and average them pairwise, the resulting 100-sample random distribution will have unit variance (per P7.13), and according to the central limit theorem, it will be Gaussian distributed.

Using the random numbers (r_i's) derived in this manner, a Poisson- and Gaussian-distributed random variable x_i can be formulated based on the following equation:

$$x_i = N + r_i. \tag{7.19}$$

The x_i's will have a mean value of N and a standard deviation of $\sqrt{N}$. This can be used to simulate a counting experiment with mean = N and standard deviation = $N^{1/2}$. This algorithm is useful in modeling photon counting for x-ray and nuclear imaging.

7.5 PROPAGATION OF ERRORS

If the random variable $z = f(x, y)$ is a function of "independent" random variables x and y, with variances σ_x^2 and σ_y^2, then the variance σ_z^2 of z is calculated as follows:

$$\sigma_z^2 = \left(\frac{\partial f}{\partial x}\right)^2 \sigma_x^2 + \left(\frac{\partial f}{\partial y}\right)^2 \sigma_y^2. \tag{7.20}$$

This can be extended to calculate the variance of any function of any number of independent random variables, if the functions and variances are known. Note that Equation 7.20 predicts Equation P.5, but without the covariance term, since it only deals with independent random variables. If covariance is not zero, then Equation 7.20 takes on the following form:

$$\sigma_z^2 = \left(\frac{\partial f}{\partial x}\right)^2 \sigma_x^2 + \left(\frac{\partial f}{\partial y}\right)^2 \sigma_y^2 + 2\left(\frac{\partial f}{\partial x}\right)\left(\frac{\partial f}{\partial y}\right) \mathrm{CoV}\{x,y\}. \tag{7.21}$$

The propagation of variance formulas should be considered as estimators of variance, since they are derived using only the linear term in a series expansion of $f(x, y)$. However, these formulas have proven adequate for many medical imaging applications.

Example 7.1: Attenuation Coefficient

We are to determine the linear attenuation coefficient of aluminum at the gamma energy of 60 keV. To do this, we have a source of ^{241}Am (emitting a 60 keV gamma) that we have collimated into a narrow pencil beam. We are counting with a 100% efficient NaI(Tl) scintillation crystal connected to a counting system. We will ignore effects due to scattered radiation or count-rate-dependent phenomenon. In other words, we assume that we have a perfect counting system. When nothing (except air) is in the beam, we obtain 6832 counts. We then place a 1 mm thick piece of pure aluminum in the beam and record 6335 counts.

(a) If there is no imprecision in the measured thickness of aluminum, what is the linear attenuation coefficient of aluminum and what is the uncertainty due to the count data given earlier?

(b) The tabulated value of the linear attenuation coefficient of aluminum is 0.7441 cm^{-1}. Is the difference between the calculated value and the tabulated value consistent with the precision you expect?

Solution

(a) We know that for narrow beam geometry, the relationship between the number of photons without the attenuator (N_0) and the number transmitted through the attenuator (N) is given by

$$N = N_0 e^{-\mu x} \tag{7.22}$$

and solving for the linear attenuation coefficients gives

$$\mu = \frac{1}{x}\ln\left(\frac{N_0}{N}\right) = \frac{1}{x}\left[\ln(N_0) - \ln(N)\right]. \tag{7.23}$$

Therefore, the calculated value of the linear attenuation coefficient is

$$\mu = \frac{1}{0.1\,\text{cm}}\ln\left(\frac{6832}{6335}\right) = 0.7553\,\text{cm}^{-1}. \tag{7.24}$$

Since N and N_0 are independent random values, the uncertainty in the calculated value is given by

$$\sigma_\mu^2 = \left(\frac{\partial \mu}{\partial N}\right)^2 \sigma_N^2 + \left(\frac{\partial \mu}{\partial N_0}\right)^2 \sigma_{N_0}^2. \tag{7.25}$$

Both gamma emission and attenuation are Poisson processes. Therefore, the variance of each measurement is equal to its mean so that

$$\sigma_N^2 = N, \quad \sigma_{N_0}^2 = N_0, \quad \frac{\partial \mu}{\partial N} = \frac{-1}{xN}, \quad \frac{\partial \mu}{\partial N_0} = \frac{1}{xN_0}, \tag{7.26}$$

leading to

$$\sigma_\mu^2 = \left(\frac{\partial \mu}{\partial N}\right)^2 \sigma_N^2 + \left(\frac{\partial \mu}{\partial N_0}\right)^2 \sigma_{N_0}^2 = \left(\frac{1}{x^2 N^2}\right) N + \left(\frac{1}{x^2 N_0^2}\right) N_0 = \frac{1}{x^2}\left(\frac{1}{N} + \frac{1}{N_0}\right). \tag{7.27}$$

Substituting $x = 0.1$ cm, $N = 6335$, and $N_0 = 6832$ into Equation 7.27, we obtain the variance and standard deviation of μ introduced by imprecision in the count data as

$$\sigma_\mu^2 = 3.042 \times 10^{-2} \text{ cm}^{-1} \quad \text{and} \quad \sigma_\mu = 0.1744 \text{ cm}^{-1}. \tag{7.28}$$

(b) The calculated value of 0.7553 cm^{-1} is within 1 standard deviation of the tabulated value of 0.7441 cm^{-1}, so the difference can be accounted for by statistical imprecision.

Example 7.2: 1-D Filter

What is the effect of using a 1-D moving average filter on a series of random values (x_i's) where the mean value of x and its variance are not changing? Let y be the output and the filter weights be $w_1 = 1/4$, $w_2 = 1/2$, $w_3 = 1/4$.

Solution

The filter is applied such that $y_2 = w_1 \cdot x_1 + w_2 \cdot x_2 + w_3 \cdot x_3 = 1/4 \cdot x_1 + 1/2 \cdot x_2 + 1/4 \cdot x_3$, where the three sequential values of x are distinguished using subscripts. The output values (y's) will be stored in a different array from the input. It follows from Equation P.1 that the mean value of y in the neighborhood of x_2 is

$$\langle y \rangle = 1/4 \langle x_1 \rangle + 1/2 \langle x_2 \rangle + 1/4 \langle x_3 \rangle.$$

Since the mean value of x is not changing with position and the weights sum to unity, this simplifies such that the mean of the output of the filter is equal to the mean of the input, $\langle y \rangle = \langle x \rangle$. The variance of y, according to Equation P.2, is

$$V\{y\} = (1/4)^2 V\{x_1\} + (1/2)^2 V\{x_2\} + (1/4)^2 V\{x_3\}.$$

And since $V\{x\}$ is consistent across x, this simplifies to

$$V\{y\} = 3/8 \cdot V\{x\}.$$

So, in this example, the mean value of x is preserved and the variance is reduced by more than 1/2, which is the objective of a smoothing filter. Note that the data in the input x series were assumed to be independent random variables. However, this is not true of output series y, since the covariance of the filtered data is nonzero, because adjacent values of x were combined to calculate y values. This property will be important in the Chapter 9 dealing with autocorrelation functions.

Example 7.3: Common Formula for Propagation of Errors

Here, z is a random variable formed from "*independent*" random variables x and y. Addition/subtraction:

$$z = x \pm y \qquad \left(\frac{\partial z}{\partial x}\right)^2 = (1)^2 \left(\frac{\partial z}{\partial y}\right)^2 = (\pm 1)^2 = 1 \quad \therefore \sigma_z^2 = \sigma_x^2 + \sigma_y^2. \tag{7.29}$$

Multiplication:

$$z = xy \qquad \left(\frac{\partial z}{\partial x}\right)^2 = y^2 \left(\frac{\partial z}{\partial y}\right)^2 = x^2 \quad \therefore \sigma_z^2 = y^2 \sigma_x^2 + x^2 \sigma_y^2$$

$$\text{and} \quad \left(\frac{\sigma_z}{z}\right)^2 = \left(\frac{\sigma_x}{x}\right)^2 + \left(\frac{\sigma_y}{y}\right)^2. \tag{7.30}$$

Division:

$$z = x/y \qquad \left(\frac{\partial z}{\partial x}\right)^2 = \left(\frac{1}{y}\right)^2 \left(\frac{\partial z}{\partial y}\right)^2 = \left(\frac{-x}{y^2}\right)^2 \quad \therefore \sigma_z^2 = \frac{1}{y^2}\sigma_x^2 + \frac{x^2}{y^4}\sigma_y^2$$

$$\text{and} \quad \left(\frac{\sigma_z}{z}\right)^2 = \left(\frac{\sigma_x}{x}\right)^2 + \left(\frac{\sigma_y}{y}\right)^2. \tag{7.31}$$

Exponentiation:

$$z = x^a \qquad \left(\frac{\partial z}{\partial x}\right)^2 = \left(ax^{a-1}\right)^2 \quad \therefore \sigma_z^2 = \left(ax^{a-1}\right)^2 \sigma_x^2 \quad \text{and} \quad \left(\frac{\sigma_z}{z}\right)^2 = a^2 \left(\frac{\sigma_x}{x}\right)^2. \tag{7.32}$$

Summary

- Net variance is the sum of the variances of each random variable whether we are adding or subtracting independent random variables.

- Net relative variance is the sum of net relative variance of each random variable whether we are multiplying or dividing independent random variables.

- In exponentiation, *relative* variance is scaled by the square of the exponent.

7.6 TRANSFORMING PROBABILITY DENSITY FUNCTIONS

The formulas for propagation of error described in Section 7.5 are useful for calculating variance for some mappings of one or more random variables (x, y, ...) to a third random variable (z). However, there are important cases when a more general approach is needed. These cases arise when the probability density function $pdf_x(x)$ of one random variable x is known, and you need to calculate $pdf_y(y)$ given the mapping function $y = f(x)$. When we determine $pdf_y(y)$, the mean and variance of y as well as higher moments can be calculated using standard formula.

The basic scheme for calculating $pdf_y(y)$ in terms of a functionally related $pdf_x(x)$ is conservation of probability such that the probability$\{y_1 \leq y \leq y_2\}$ = probability$\{x_1 \leq x \leq x_2\}$, where $y_1 = f(x_1)$ and $y_2 = f(x_2)$. This conservation of probability is stated mathematically as follows:

$$\int_{y1}^{y2} pdf_y(y)dy = \int_{x1}^{x2} pdf_x(x)dx. \tag{7.33}$$

Note the use of subscripts for the pdf's to indicate which is for x and which is for y. Also recall that the integral over the full range of each pdf is unity. Importantly, Equation 7.33 assumes that the mapping function $y = f(x)$ is monotonic so that a one-to-one mapping of x to y occurs. This holds for simple linear mapping functions such as $y = ax + b$. However, and for an important case such as $y = x^2$, Equation 7.33 can be modified to separately cover positive and negative ranges as follows:

$$\int_{y1}^{y2} pdf(y)dy = \int_{x1}^{x2} pdf(x)dx + \int_{-x2}^{-x1} pdf(x)dx. \tag{7.34}$$

For pdf's that are symmetric about $x = 0$, both integrals have the same area, and this leads to a simpler equation for this case:

$$\int_{y1}^{y2} pdf(y)dy = 2\int_{x1}^{x2} pdf(x)dx. \tag{7.35}$$

Higher-order polynomials and periodic functions also follow this multi-integration range scheme though we will not deal with those here.

If we look at a limiting case of Equation 7.33 where the ranges of integrations are reduced to dy and dx, then we see that

$$\text{pdf}_y(y)dy = \text{pdf}_x(x)dx$$

$$\text{pdf}_y(y) = \text{pdf}_x(x)\frac{dx}{dy}.$$ (7.36)

Rearranging Equation 7.36 and using the absolute value of the derivative leads to

$$\text{pdf}_y(y) = \frac{\text{pdf}_x(x)}{|dy/dx|}.$$ (7.37)

The absolute value is needed, since dy may be increasing or decreasing as dx increases but the areas associated with the pdf's are always positive. We want everything on the right side of Equation 7.37 to be in terms of y and we can accomplish this by expressing x as $f^{-1}(y)$:

$$\text{pdf}_y(y) = \frac{\text{pdf}_x(f^{-1}(y))}{|dy/dx|} = \frac{\text{pdf}_x[x(y)]}{|dy/dx|}.$$ (7.38)

This assumes that $f^{-1}(y)$ exists, which will be true for the pdf's we will study. Equation 7.38 shows that $\text{pdf}_y(y)$ can be calculated by replacing the argument of $\text{pdf}_x(x)$ with $x = f^{-1}(y)$ and dividing by the absolute value of dy/dx. Therefore, $\text{pdf}_x()$ serves as the starting template, which is scaled by $1/|dy/dx|$ to calculate $\text{pdf}_y()$.

For students interested in how this technique can be expanded to deal with more complex functional relationships, look for chapters on *Probability and Functions* in classical statistics texts such as the one by Papoulis.

Example 7.4: Shifting and Scaling the Normal Distribution

Given that x is a standard (normal) Gaussian random variable ($\mu_x = 0$, $\sigma_x = 1$) and that $y = ax + b$, what is $\text{pdf}_y(y)$?

Given: $\text{pdf}_x(x) = \dfrac{1}{\sqrt{2\pi}}e^{-\frac{1}{2}x^2}$,

$f^{-1}(y) = x(y) = (y - b)/a$ and $|dy/dx| = |a|$, so using Equation 7.38, we see that $\text{pdf}_y(y)$ is

$$\text{pdf}_y(y) = \frac{1}{\sqrt{2\pi}\,|a|}e^{-\frac{1}{2}\left(\frac{y-b}{a}\right)^2} = \frac{1}{\sqrt{2\pi}\sigma_y}e^{-\frac{1}{2}\left(\frac{y-\mu_y}{\sigma_y}\right)^2}.$$

The resulting $\text{pdf}_y(y)$ is also a Gaussian function but with $\mu_y = b$ and $\sigma_y = |a|$. The transform $y = ax + b$ when applied to the Gaussian-distributed random variable x with ($\mu_x = 0$, $\sigma_x = 1$) produces a Gaussian-distributed random variable y with ($\mu_y = b$, $\sigma_y = |a|$). This simple linear function can therefore be used to formulate y's with any desired μ_y and σ_y values.

Example 7.5: Transforming Gaussian to Normal Distribution

A common use of this simple linear function is to transform a Gaussian random variable with known mean and standard deviation to a normal (Gaussian) random variable ($\mu = 0$, $\sigma = 1$). Transforming a Gaussian-distributed random variable x to a normally distributed random variable z is done using the function $z = (x - \mu_x)/\sigma_x$ where μ_x and σ_x are the mean and standard deviation for the random variable x.

Given: $\text{pdf}_x(x) = \dfrac{1}{\sqrt{2\pi}\sigma_x}e^{-\frac{1}{2}\left(\frac{x-\mu_x}{\sigma_x}\right)^2}.$

Calculate: $f^{-1}(z) = x(z) = \sigma_x z + \mu_x.$

$\qquad\quad |dz/dx| = 1/\sigma_x.$

Using Equation 7.38 with z in place of y, we see that $\text{pdf}_z(z)$ is

$$\text{pdf}_z(z) = \frac{1}{\sqrt{2\pi}}e^{-\frac{1}{2}z^2},$$

which is the pdf for normal random variable z. Probability tables are available in most statistical texts for this normal z distribution, so probabilities for x between any range can be calculated by determining the corresponding range of z in the table.

Random number generators for z can be found in various software packages (Mathcad, MATLAB®, Mathematica, etc.). The scaling transform $x_i = \sigma_x z_i + \mu_x$ as shown in Example 7.1 can be used to generate Gaussian random variables with mean $= \mu_x$ and standard deviation $= \sigma_x$.

Example 7.6: Nonlinear Transformation of Gaussian Distribution

Given that x is a zero-mean Gaussian random variable with variance of σ_x^2 and that $y = ax^2$ ($a > 0$), what is $\text{pdf}_y(y)$? (Note that $y \geq 0$)

Given: $\text{pdf}_x(x) = \dfrac{1}{\sqrt{2\pi}\sigma_x} e^{-\frac{1}{2}\left(\frac{x}{\sigma_x}\right)^2}$.

Calculate: $f^{-1}(y) = x(y) = \pm\sqrt{\dfrac{y}{a}}$

$$|dy/dx| = |2ax| = \left|2\sqrt{ay}\right|.$$

Solving for $\text{pdf}_y(y)$ makes use of Equation 7.35, since there are two values of x for each value of y. Using Equation 7.38 with this modification, this leads to

$$\text{pdf}_y(y) = \frac{1}{\sqrt{2\pi a y}\,\sigma_x} e^{-\frac{1}{2}\frac{y}{a\sigma_x^2}}$$

Note that this $\text{pdf}_y(y)$ is no longer a Gaussian distribution, as should have been expected, since the transform function was nonlinear and particularly since $y > 0$ though x can be positive or negative. The new pdf has the form of a Gamma distribution function.

7.7 GRAPHICAL ILLUSTRATIONS OF TRANSFORMS OF RANDOM VARIABLES

The basic principle in transforming random variable x to random variable y is based on matching probabilities. Probabilities can be determined from the cumulative distribution functions [$\text{cdf}_y(y)$ and $\text{cdf}_x(x)$]. The general scheme is that y corresponds to x where $\text{cdf}(y) = \text{cdf}(x)$. As such, the matching probabilities have the same vertical extent in the graph of Figure 7.3. Figure 7.3 shows how to convert between two pdf's given their cdf's. The formulas for this are

$$y = \text{cdf}_y^{-1}\left(\text{cdf}_x(x)\right),$$

$$x = \text{cdf}_x^{-1}\left(\text{cdf}_y(y)\right).$$

This approach can be used to the convert gray level histogram of one image to match that of another similar image, which can correct for minor differences in contrast.

Figure 7.4 illustrates how to convert from a Gaussian random variable x to a uniform random variable y. In this example, the value of $x = 20$ is transformed to the value of $y = -10$. We see that probabilities match when $\text{cdf}(y)$, calculated as $(y - y_{min})/(y_{max} - y_{min})$, is equal to $\text{cdf}_x(x)$. The y_{range} for this example is 100. The range of x does not have to match

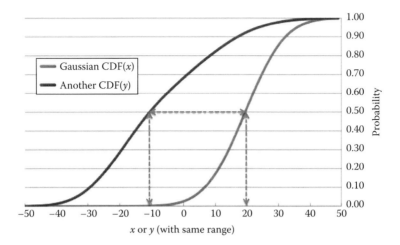

FIGURE 7.3 Graphical illustration of conversion between a Gaussian-distributed random variable x and another random variable y with the same cumulative probability using their cumulative distribution functions (CDFs).

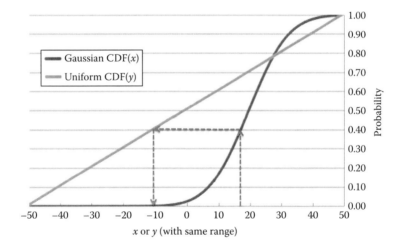

FIGURE 7.4 An example for converting from a Gaussian-distributed random variable x ($\mu = 20$, $\sigma = 10$) to a uniform distributed random variable y ($y_{min} = -50$, $y_{max} = 50$).

that of y but we forced them to fall in the same range for this example so that both could be plotted on the same graph. This approach has been used for histogram equalization of medical images.

The conversions illustrated in Figures 7.3 and 7.4 can be implemented using mathematical software applications such as Matlab, Mathcad, and Mathematica as well as some image-processing applications. One use is the conversion from t-distributed random variables to z-distributed random values, since probabilities are well documented for z-values.

HOMEWORK PROBLEMS

P7.1 A measurement is made of the time interval between successive disintegrations of a radioactive sample. After a large number of measurements, we determine that the disintegrations occur completely at random with a mean rate (θ). We find that the time interval between successive disintegrations follows an *exponential distribution* function $f(t)$, given by

$$f(t) = \theta e^{-\theta t}.$$

(a) Show that $f(t)$ behaves like a normalized distribution function over the domain $[0, +\infty)$ by proving that $\int_0^{+\infty} f(t)dt = 1$.

(b) Prove (by direct calculation) that the mean time interval of the exponential distribution is $\mu = 1/\theta$ and the standard deviation is $\sigma = 1/\theta$.

(c) If we perform an experiment and find that the mean time interval between successive disintegration is 3 s, find the probability that a disintegration will occur within 2 s of a preceding one.

P7.2 Write a computer program in any language (Matlab, C, Basic, Java, etc.) or with a spreadsheet to compute 100 Poisson-distributed numbers with an arithmetic mean equal to 61. (*Extra credit*: Determine the Poisson distribution for this example and compare it against the distribution of numbers that you obtained from your computer program.)

P7.3 You perform a counting experiment with a weak radioisotope source and a Geiger counter in which you measure 269 counts in 1 min.

(a) Derive the formula for the Poisson probability distribution that describes the probability of obtaining N counts in 1 s (not 1 min!).

(b) Assuming that the radioisotope source has a relatively long half-life (so that the count rate does not change during your measurement), what is the probability of measuring 3 or more counts in any 1 s time interval?

P7.4 You have assigned a student helper to make counting measurements on a radiolabeled protein and instructed him to repeat the measurement 10 times so that you can calculate the mean and standard deviation. The student works on this project and when he returns, he reports that he has made 10 measurements with the following count values:

105, 104, 107, 102, 101, 106, 105, 107, 107, 108.

(a) Calculate the mean and standard deviation of these measurements, and show that the measured standard deviation is smaller than that you would expect from a counting experiment.

(b) Is the observation in (a) possible for a radiation counting experiment? Discuss possible reasons why the measured standard deviation could be *smaller* than the theoretically predicted value?

P7.5 The exponential pdf $f(t) = \theta e^{\theta t}$ describes how successive disintegrations are distributed in time.

(a) Given that the minimum resolving time of a detector is 1 ms, at what true count rates will (i) 10%, (ii) 20%, and (iii) 50% of the true events be lost.

(b) What is the observed count rate for each case?

P7.6 The distance between two points in a 3-D image is calculated as

$$d = \sqrt{\left(x_1 - x_2\right)^2 + \left(y_1 - y_2\right)^2 + \left(z_1 - z_2\right)^2},$$

where (x_1, y_1, z_1) and (x_2, y_2, z_2) are the coordinates of the two points and $(\sigma_{x1}, \sigma_{y1}, \sigma_{z1})$ and $(\sigma_{x2}, \sigma_{y2}, \sigma_{z2})$ are associated standard deviations. Calculate the standard deviation in the calculated distance (σ_d) in terms of these parameters using propagation of errors.

P7.7 For Example 7.1, what thickness of aluminum attenuator would provide the smallest value for the variance of the measured linear attenuation coefficient? Determine an approximate relationship that would allow you to estimate the optimal attenuator thickness for material with different linear attenuation coefficients.

CHAPTER 8

Noise and Detective Quantum Efficiency

8.1 INTRODUCTION

Noise is generally defined as the uncertainty in a signal due to random fluctuations in the signal. There are many causes for these fluctuations. For example, the fluctuation in x-ray beam intensity emerging from an x-ray tube is naturally random. In fact, the number of photons emitted from the source per unit time varies according to a Poisson distribution. Other sources of random fluctuation that are introduced by the process of attenuation of the materials present in the path of the radiation beam (patient, x-ray beam filtration, patient table, film holder, and detector enclosure) are also Poisson processes. Finally, the detectors themselves often introduce noise. In a film-screen cassette, both the intensifying screen and the film contain individual grains that are sensitive to the radiation or light. Therefore, the exposure of the grains in the film produces random variations in film density on a microscopic level, which is a source of noise. When electronic detectors are used, the absorption of radiation by the detector is a random process as well as inherent electronic noise. Electronic detectors generate currents from thermal sources that introduce random fluctuations into the signal. RF and other sources of noise are present in MR images. As such, we see that noise is inherent in all medical imaging systems, so it is important to investigate and characterize the sources of noise.

8.2 SIGNIFICANCE OF NOISE IN MEDICAL IMAGING

The importance of noise in medical x-ray imaging arises because x-rays are ionizing and can damage molecules of biological importance such as DNA or can cause cell death. We therefore seek to minimize patient exposure, and produce medical radiographs that are intrinsically "noise-limited." That is, at the exposure levels obtained in radiographs, our ability to discern objects of interest may be limited by the presence of noise rather than, for example, limitations in spatial resolution. Indeed, on an ethical basis when ionizing radiation is used to form a medical image, we are compelled to minimize patient exposure so that the image is *quantum-limited* to a degree that allows us to derive essential information and make the best possible diagnosis. If this is not the case, then we are potentially delivering additional exposure to the patient without the benefit of additional diagnostic information. Alternatively, if our image is limited by electronic noise or noise sources other than photons, then we have not designed our imaging system correctly, since we should not let noise sources other than fundamental unavoidable photon statistical noise interfere with our ability to derive information from medical images.

8.3 DESCRIPTIVE QUANTITATION OF NOISE

There are various ways to quantify the level of noise in an image. In the previous chapter, we discussed random processes for which we knew the underlying probability distributions. In experimental studies, we may not know the exact nature of the probability distribution that describes the random process. Rather, we have a noisy signal, which, in an experimental setting, we can measure as many times as we wish. After a number of measurements, we calculate the sample mean and the standard deviation using well-known formulae. If x_i's are individual measurements, then the sample mean (m) and sample variance (s^2) are given by

$$m = \langle x_i \rangle = \frac{1}{N} \sum_{i=1}^{N} x_i \quad \text{and} \quad s^2 = \left\langle \left(x_i - m \right)^2 \right\rangle = \frac{1}{N-1} \sum_{i=1}^{N} \left(x_i - m \right)^2, \qquad (8.1)$$

where N is the number of measurements. Here, the brackets <> represent the averaging operation. The square root of sample variance s is the sample standard deviation, which we can use to quantify the uncertainty, or noise associated with a signal. The division by $N - 1$ rather than N to estimate variance (Equation 8.1) makes s^2 an unbiased estimate of the population variance (σ^2).

Theoretically, in medical imaging, the calculation of the mean and standard deviation would require multiple measurements (or multiple images) on the same object. For example, if one wanted to know the noise in a radiograph of a lung nodule from a patient, a large number of radiographs taken under identical conditions would have to be obtained. The restriction of identical condition is impossible to achieve for a patient because of tissue movement due to the heart beating or the patient breathing could (and probably would) move the tumor so that identical measurements could not be made from repetitive radiographs. But assuming that these difficulties could be overcome, and we did obtain multiple radiographs under identical patient conditions, the measurement of noise could be made by measuring x_i at the same location in the radiograph from all of the radiographs. From these measurements, m and s could be calculated, giving the average and the uncertainty of the measured value but just at that single point in the image. To fully characterize the noise properties of the image, this process would have to be repeated for each location in the image.

Of course, this is rarely done except perhaps by a maniacal graduate student working on his or her dissertation and making measurement on a phantom rather that a patient.

Usually, to characterize the noise in an image, we obtain a single image that has a large region where the signal is uniform, and measure the mean and variance using multiple samples within this region. In radiography, this can be achieved by imaging a block of acrylic or some other uniformly thick material. In each case, the object placed in the x-ray beam mimicking a person is called a phantom. It is also a common practice to have nothing in the x-ray beam (except for air—hence the term "air-scan") to obtain an image for such noise measurements. In each case, we assume that the signal is the same (except for random fluctuations) at every location within the region of interest, and calculate the noise (standard deviation) using measurements at multiple sites within the region. This provides a good estimate of quantum noise assuming that there are no spatially correlated noise sources within the region of interest. A spatially correlated noise source might be generated by an oscillation in a video amplifier, a dirty roller in the film processor, or by some other (possibly unknown) process that affects neighboring values in the region of interest.

8.3.1 Signal-to-Noise Ratio (SNR)

The standard deviation is the most useful way to quantify noise in an imaging system. However, for the description of noise to have practical meaning, it needs to be evaluated relative to signal size. Since the noise specifies the uncertainty in the signal, it is important to relate noise size to signal size. For example, an electronic signal that has a noise magnitude of 2 mV is 10% of a 20 mV signal, but only 0.01% of a 20 V signal. For this reason, the concept of SNR is used to describe the relationship between signal and noise magnitude (standard deviation). The signal and noise must be measured in the same units, relative exposure for a film image, electric potential (volts) for an electronic image, or photon fluence (or exposure, etc.) for a radiographic signal. Once both signal and noise magnitude are determined, simple division of the mean signal by the signal standard deviation is used to calculate the SNR.

8.3.2 Detective Quantum Efficiency (DQE)

An imaging device that is perfect in terms of noise performance is one that does not add noise, that is, does not degrade the SNR of the input signal. It is difficult, if not impossible, to improve SNR without degrading some other aspect of system performance (e.g., spatial or temporal resolution). How the system affects the SNR is an important characteristic quantified as the DQE of the system.

By way of definition, if an instrument or device receives information or data with an SNR of SNR_{in}, from which it produces information or data with an SNR of SNR_{out}, then the DQE of the instrument or device is

$$DQE = \left(\frac{SNR_{out}}{SNR_{in}} \right)^2. \tag{8.2}$$

A perfect device (in terms of SNR) is one that maintains an SNR of all signals presented to it and would therefore have DQE = 1. Since SNRs are unitless, input and output signals do not have to be measured using the same units when calculating DQE. For example, the input signal might be measured using count rate and the output signal measured using voltage. In the following example, you will see why it is useful to square the SNRs in the calculation of DQE.

Example 8.1: DQE of NaI(Tl) Detector

Assume we use a sodium iodide detector of thickness a with a counting system to measure gammas from a ^{99m}Tc source (140 keV gammas). Assume also that the counting system generates a signal equal to the number of photons detected, and that the only noise source is due to the Poisson statistics of the detected photons. Can you think of other sources of random noise in such a system?

If the linear attenuation coefficient of the sodium iodide at this energy is μ, and all gammas interacting with the sodium iodide crystal are detected, what is the DQE of this detector?

Solution

If N_0 gamma ray photons are incident on the sodium iodide detector, then the number of photons that are transmitted through the detector without detection is

$$N_t = N_0 e^{-\mu a} \tag{8.3}$$

and the number of photons counted by the detector is

$$N = N_0 - N_t = N_0\left(1 - e^{-\mu a}\right). \tag{8.4}$$

The noise in the signal input to the detector is due to the statistical fluctuation in the number of photons incident on the detector. Therefore,

$$(\text{SNR})_{\text{in}} = \frac{\text{Signal}}{\text{Noise}} = \frac{N_0}{\sqrt{N_0}} = \sqrt{N_0}, \tag{8.5}$$

while the SNR due to the detector is determined from the photons counted:

$$(\text{SNR})_{\text{out}} = \frac{\text{Signal}}{\text{Noise}} = \frac{N_0\left(1 - e^{-\mu a}\right)}{\sqrt{N_0\left(1 - e^{-\mu a}\right)}} = \sqrt{N_0\left(1 - e^{-\mu a}\right)}. \tag{8.6}$$

Therefore, the DQE of the detector system is

$$\text{DQE} = \left(\frac{\text{SNR}_{\text{out}}}{\text{SNR}_{\text{in}}}\right)^2 = 1 - e^{-\mu a}. \tag{8.7}$$

Note that as the product μa becomes very large, DQE approaches unity as the detector absorbs most gammas (i.e., the detector becomes "near perfect" in terms of preserving the SNR). Alternatively, if the value of μa is small, then $e^{-\mu a}$ can be expanded in a Taylor's series and neglecting higher-order terms simplifies DQE further

$$e^{-\mu a} \approx 1 - \mu a \tag{8.8}$$

so that

$$\mathrm{DQE} \approx \mu a. \tag{8.9}$$

The DQE approaches zero in the limit of a low photon efficiency (small μ) and/or thin detector (small a).

8.4 OPTIMIZATION OF RADIOGRAPHIC SYSTEM SNR

There are several procedures that can be utilized to improve the SNR in an image (Table 8.1). Generally, but not always, these procedures sacrifice some other aspect of the medical image. *It is one role of the physicist or engineer to balance these various requirements in order to obtain maximal information from the image.* For example, if we double the resolution width of a 2-D detector, this will improve our SNR two fold (the number of photons collected is proportional to the detector area, so increases four fold). However, this improvement in the noise characteristics is accompanied by a decrease in detector's spatial resolution. This may or may not be a problem depending on the spatial resolution needs and limits imposed by other components in the system. Similarly, one can increase the x-ray tube current (mA) or exposure time, allowing more photons to be produced and detected. However, increasing mA may require a larger focal spot size, decreasing detail, and increasing exposure time may lead to additional motion unsharpness, so these approaches need to be carefully considered.

8.4.1 Optimization of Photon Energy

The selection of x-ray photon energy is important in establishing the SNR for an imaging task. The basic issue with this optimization is that photon energy forces a trade-off between increasing radiation dose at lower energies and decreasing contrast at higher energies.

TABLE 8.1 Ways to Increase the Quanta SNR for X-Ray Systems

Method	Disadvantages
Increase detector size (area)	Degrades spatial resolution
Increase tube current	Increases patient exposure
Increase exposure time	Increases patient exposure and motion
Increase kVp (mean energy)	Decreases radiographic (subject) contrast
Increase detector thickness	May degrade spatial resolution (screens)

The trade-off between dose and SNR is tedious to analyze for several reasons. First, most attenuators, including simple ones such as water, contain more than one element, requiring estimation of the linear attenuation coefficients from the elemental constituents. Second, linear attenuation coefficients are complex functions of both energy and atomic number. Third, the polyenergetic nature of the x-ray beam forces evaluation of integrals for determining detector response. Generally, the integrals cannot be evaluated in closed form, requiring numerical methods.

However, it is instructive to estimate the SNR per unit radiation dose as a function of x-ray beam energy and determine if there is an optimal energy. In this estimate, we consider the SNR of a 1 cm thick void region within a volume of water having thickness x (Figure 8.1), and the photon beam is considered to be monoenergetic with an energy E.

First, we calculate the SNR of the radiographic signal. We will use the following object and x-ray system parameters for this:

x, the thickness of water region

φ_0, the incident photon fluence (actually exit fluence w/o phantom)

φ_1, the photon fluence through x cm of water (number/area)

φ_2, the photon fluence through 1 cm air void and $x - 1$ cm of water

μ, the linear attenuation coefficient of water

E, the photon energy

ρ, the density of water

A, the cross-sectional area of interest

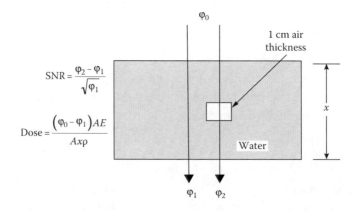

FIGURE 8.1 Geometry for calculating SNR and radiation dose and as a function of x-ray beam energy for a 1 cm thick void within a volume of water having thickness x. The photon beam is monoenergetic with photon energy E.

The photon fluence φ_2 in the region of interest is

$$\varphi_2 = \varphi_0 e^{-\mu(x-1)} \tag{8.10}$$

and the photon fluence φ_1 in the background is

$$\varphi_1 = \varphi_0 e^{-\mu x}. \tag{8.11}$$

The signal is the difference in photon fluence between these two regions, and the noise will be estimated as the standard deviation (square root) of the background photon fluence. *Note: In the low contrast case $\varphi_1 \sim \varphi_2$, so using the background alone to assess noise is justified, and noise is easier to measure in the larger background region.* Therefore, the SNR for this simple model is

$$\text{SNR} = \frac{\varphi_2 - \varphi_1}{\sqrt{\varphi_1}}. \tag{8.12}$$

The mean radiation absorbed dose is estimated as the energy absorbed in the water for a radiation beam of cross-sectional area A divided by the mass of the water in the beam (Equation 8.13). We will approximate the absorbed energy as the number of photons, $(\varphi_0 - \varphi_1)A$, multiplied by the photon energy E, limiting to the case of no scattering. The mass in the beam is the product of beam area A, beam length x, and water density ρ. This leads to an estimated mean absorbed dose of

$$\text{Mean absorbed dose} = \frac{(\varphi_0 - \varphi_1)AE}{Ax\rho} = \frac{(\varphi_0 - \varphi_1)E}{x\rho}. \tag{8.13}$$

Note that the estimated mean absorbed dose is not a function of the area A. The SNR and the mean absorbed dose are graphed as a function of photon energy E in Figure 8.2.

Figure 8.3 shows that the optimal response (i.e., maximum SNR per dose) decreases with increasing thickness. This model suggests that a higher optimal SNR per dose is possible for thinner body parts and that the optimal SNR decreases substantially for larger bodies and parts, potentially leading to lower quality studies. Additionally, the optimum SNR for larger bodies requires higher photon energy.

8.4.2 Selection of Detector Material

Another important consideration in limiting quantum statistical noise is selection of appropriate material to maximize photon absorption by the detector. Increasing the thickness as well as the linear attenuation coefficient of the phosphor material can increase absorption. As we showed in the chapter on spatial resolution (Equation 6.28), increasing

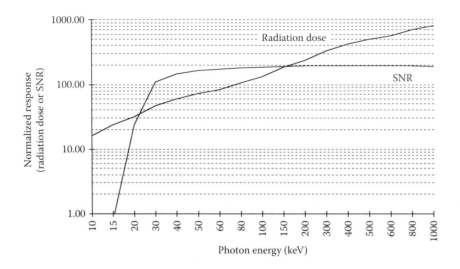

FIGURE 8.2   Based on the model in Figure 8.1 and assuming a fixed number of photons entering a volume of water containing a 1 cm air void, both the radiation dose and the signal-to-noise increase as a function of photon energy for $E < 30$ keV.

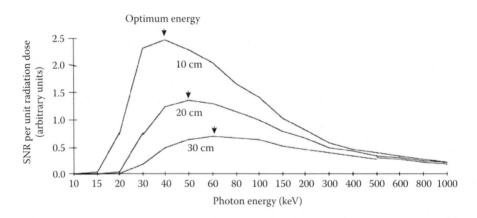

FIGURE 8.3   The SNR per unit radiation dose based on the model in Figure 8.1 varies considerably for both object thickness and photon energy.

the phosphor thickness increases geometric unsharpness, suggesting that increasing the linear attenuation coefficient would be the preferred approach.

The choice of phosphor material must be carefully matched with the photon energy used. For best SNR characteristics, the linear attenuation coefficient of the detector should be highest in the energy region where the transmitted x-ray spectrum is highest and, in particular, where the contrast of interest is highest.

For many years, calcium tungstate ($CaWO_4$) was the most common material used in intensifying screens. In these screens, x-ray absorption is primarily provided by tungsten, which contributes to both the physical density and a high atomic number of the phosphor. To understand whether the attenuation coefficient of tungsten is well suited as a phosphor

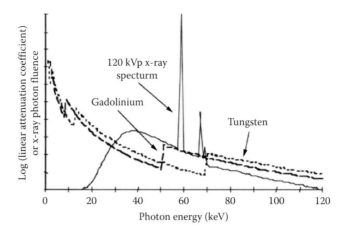

FIGURE 8.4   For a typical x-ray spectrum used in diagnostic radiology (120 kVp), a rare-earth intensifying screen (Gadolinium) will absorb a greater fraction of the x-rays than a conventional phosphor (Tungsten) for photons with energy ranging from 50 to 70 keV.

for medical imaging, we must understand how it matches with the energy spectrum of the x-ray beam. This can be appreciated using a general rule of thumb to estimate the characteristics of an x-ray spectrum. This rule states that the effective energy of a diagnostic x-ray beam is ~1/3 of its peak kilovoltage (kVp). Hence, a diagnostic examination using a 120 kVp beam should use a phosphor with its maximal attenuation at 120/3 = 40 keV. However, the K-edge of tungsten is 69.5 keV, in many cases too high to contribute to attenuation of the many of the x-ray photons for this x-ray beam (Figure 8.4). Additionally, the K-edge for calcium is too low ~4 keV.

For this reason, phosphors with rare-earth components were introduced and are widely used in diagnostic examinations. Typical phosphors include gadolinium oxysulfide (GdO_2S:Tb), lanthanum oxybromide (LaOBr:Tb), and yttrium oxysulfide (Yt_2O_2S:Tb). These phosphors have K-edges ranging from 39 keV to about 50 keV, which improves their x-ray absorption for many diagnostic examinations (see Figure 8.4 for Gd).

Just as the x-ray absorption characteristics of the phosphor must be matched to the spectrum of the x-ray beam, the spectral response of the radiographic film must be matched to the spectral light output of the intensifying screens. Gadolinium oxysulfide (GdO_2S:Tb) emits light with a maximum emission at 545 nm. It should be matched with a film sensitive to *green light* (wavelength 550 nm). Another useful phosphor material is lanthanum oxybromide (LaOBr:Tb) that emits light in the range of 380–450 nm. It should be matched to a *blue-sensitive* film, one having an emulsion sensitive to photon wavelengths shorter than 500 nm.

When phosphors are viewed in fluoroscopy systems, the spectral output of the phosphor must be matched to the sensitive region of the human eye or video system used to record the image. Before the advent of image intensifiers, fluoroscopic screens used zinc-cadmium sulfide (a mixture of ZnS and CdS), which has a spectral output well matched to the spectral response of the human eye. This material is used in the input phosphor of older image intensifiers, although it has been replaced by cesium iodide (CsI) in newer systems (*see if you know why*). Zinc-cadmium sulfide is still used as the output phosphor for image intensifiers, since its spectral emission lies in the visible range of the electromagnetic spectrum and is well matched to the spectral response of optical devices.

Finally, laser film printers utilize a helium neon laser with a red light beam to expose the film. The films used with laser printers are sensitive in the infrared region, but also can be exposed by the red light used to provide low-level illumination (i.e., the "safe light") in most dark rooms. An operator forgetting their physics will be very disappointed with their results if they forget to turn off the red light when removing a laser printer film in the dark room for placement in the film processor. Modern laser film printers have an attached film processor, so this problem has mostly resolved.

8.5 NOISE SOURCES IN THE MEDICAL IMAGE

Of the several sources of noise in the medical image, some are introduced by the chemical or photographic limitations of our technology. However, there is a fundamental and unavoidable noise source against which we are always fighting in x-ray imaging, namely, photon statistical noise or *quantum noise*. By quantum noise, we mean the statistical imprecision introduced into a radiation signal by the random fluctuations in photon production, attenuation, and recording. These are natural sources that cannot be avoided.

For a detector, the photon statistical noise is based on the number of photons recorded and used to generate the image. For example, in nuclear medicine imaging, we detect both primary and scattered gammas, but a spectrometer energy window favoring primary gammas is used to selectively ignore most scattered gammas. Photons that pass through the detector without being absorbed, or even those that are absorbed without generating image information, are wasted and do not contribute to reducing noise in the image. Both signal and noise decrease as the number of detected photons decreases, but signal more so than noise, leading to a reduction in SNR. Recall that for many cases, the SNR can be estimated as $N^{1/2}$. Since photons cannot be subdivided, they represent the fundamental quantum level of a system. In conventional radiographic imaging systems, x-ray photons and light photons are the quanta of importance. The point along the imaging chain where the fewest quanta are used to represent the image is called the "quantum sink." *The noise level at the quantum sink determines the noise limit of the entire imaging system.* Therefore, without increasing the number of information carriers (i.e., quanta) at the quantum sink, the system noise limit cannot be improved.

Example 8.2: Film-Screen Quantum Sink

A screen with 10% conversion *efficiency* is one that converts 10% of the absorbed x-ray energy to light photon energy. A 20% conversion *efficiency* screen produces twice as many light photons though with the same x-ray absorption.

In a film-screen cassette, an intensifying screen having 10% conversion *efficiency* is replaced by one having 20% conversion efficiency. Both screens have 50% x-ray

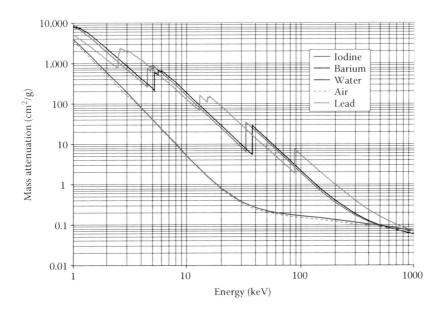

FIGURE 4.5   Both iodine and barium and K-absorption edges in the range of diagnostic x-ray imaging. Their mass attenuation coefficients are significantly larger than those of water (tissue equivalent) and air, so they are excellent contrast agents to distinguish vessels or gastrointestinal tract bounded by air and/or soft tissues.

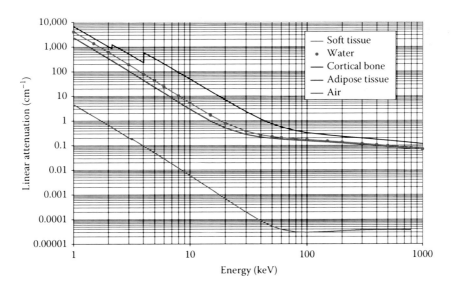

FIGURE 4.7   The linear attenuation coefficient is the product of the mass attenuation coefficient (cm²/g) and density (g/cm³). This figure shows the tremendous difference between air and other body tissues, making it an excellent contrast agent.

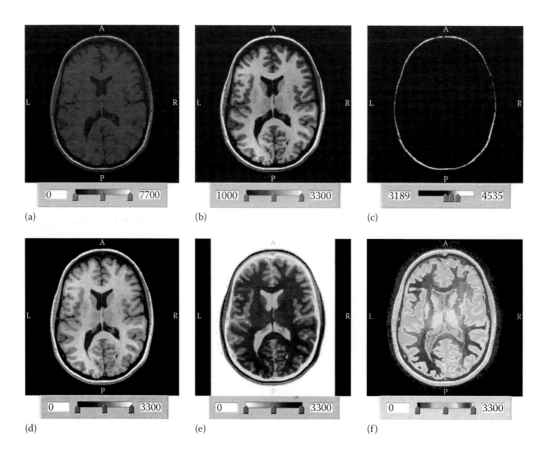

FIGURE 4.13 MRI brain image with display range settings (a) over the full range of values (0–7700), (b) for high contrast between gray and white matter (1000–3300), (c) highlighting fat tissue in the scalp (3189–4535), (d) spanning brain tissue values (0–3300), (e) range as in (d) but with a negative LUT, and (f) a color LUT can be used to distinguish tissue ranges by color.

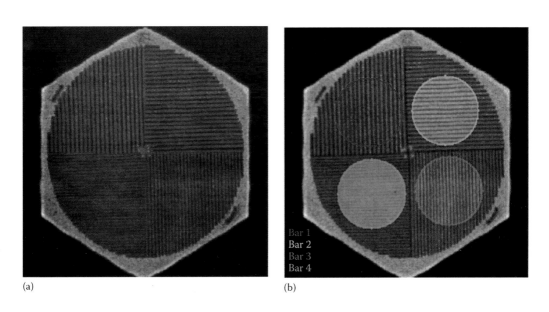

FIGURE 15.1   Bar phantom image for LFOV camera head (a) and ROI for assessing change of contrast with decreasing bar size (b).

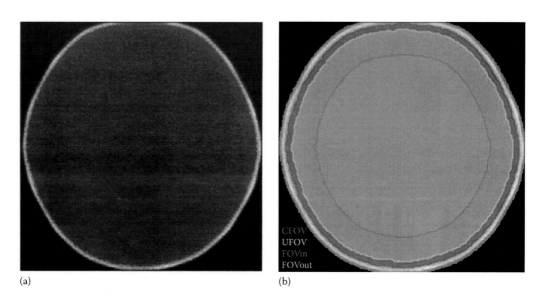

(a)   (b)

FIGURE 15.3   (a) Flood image for LFOV camera head with (b) ROIs for the FOV (outer and inter-ridge), UFOV, and CFOV.

absorption *efficiency* for 50 keV x-ray photons, and the film is assumed to absorb 100% of the light emitted by the screen.

(a) In each case, where is the quantum sink (input to screen, in the screen, or input to the film)?
(b) If the same film is used with both intensifying screens, which radiograph will be noisier?

Solution

(a) With the 10% conversion efficiency phosphor, 50,000 eV/10 = 5,000 eV of the absorbed 50 keV x-ray photon's energy is converted to light photon energy. The light photons have a wavelength of about 5000 A (Å = Angstroms) corresponding to energy of

$$E = \frac{12{,}396 \text{ eV} \cdot \text{Å}}{5{,}000 \text{ Å}} \approx 2.5 \text{ eV}, \tag{8.14}$$

so the estimated number of light photons produced per 50 keV x-ray photon absorbed is

$$N = \frac{5000 \text{ eV/x-ray photon}}{2.5 \text{ eV/light photon}} = 2000 \frac{\text{light photons}}{\text{x-ray photon}}. \tag{8.15}$$

Obviously, for a 20% conversion efficiency phosphor, this would increase to 4000 light photons/x-ray photon.

   If the input to the screen is 100 x-ray photons, the screen reduces this to 50 absorbed x-ray photons, and the film sees 100,000 (50 × 2,000) to 200,000 (50 × 4,000) light photons, depending on which screen is used. Therefore, if the film exposure is the same, then the "quantum sink" is due to the small number of x-ray photons absorbed in the intensifying screen (Figure 8.5).

(b) When considering which radiograph will be noisier, it is important to remember that light exposing the film must be maintained in the linear exposure region of the film H&D curve. Assume we obtain the correct film exposure with the 10% efficient phosphor. When the 20% conversion efficiency phosphor is used, the x-ray exposure must be halved to maintain the same photographic exposure and resulting film density. At the quantum sink (i.e., the intensifying screen), fewer (x-ray) photons are used to create the image with the 20% conversion efficiency phosphor than the 10% phosphor; however, patient exposure is reduced by 50%. In this example, the image obtained with the 20% is noisier than that obtained with the 10% conversion efficiency phosphor (Figure 8.5).

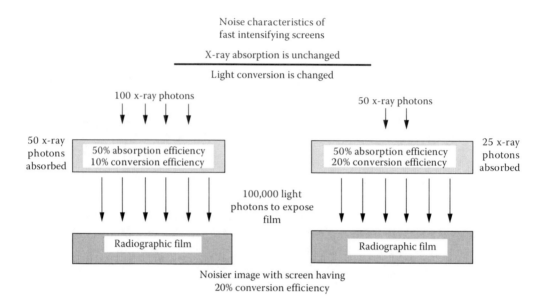

FIGURE 8.5 A more efficient phosphor uses fewer x-ray photons than a less efficient phosphor to expose radiographic film. Therefore, when using the same film, the more efficient phosphor (i.e., the fast screen) produces a noisier image than the less efficient phosphor.

Fortunately, it is relatively easy to characterize photon statistics, at least at a descriptive level. This is because photon production and attenuation are Poisson statistical processes. As pointed out before, a valuable result is that the standard deviation for the Poisson distribution equals the square root of the mean. In most cases, we can estimate the standard deviation by taking the square root of a single measurement of the number of photons. Of course, the next measurement generally will result in a different value, but this difference should be small if the variability is small compared to the signal (with a signal of $N = 10,000$, the standard deviation is just 1% of the mean). Furthermore, for $N > 20$–30, the Poisson distribution becomes Gaussian-like, which is useful for two reasons. First, the Gaussian distribution has the well-known property that 68.3% of the observations will fall within ± 1 standard deviation of the mean, 95.5% of the observations will fall within ± 2 standard deviations of the mean, and 99.7% of the observations will fall within ± 3 standard deviations of the mean. The second useful property of the Gaussian distribution is that it has a nice mathematical form that is useful in generating homework problems for graduate students!!

Example 8.3: Screen Viewing Quantum Sink

A radiologist views a conventional fluorescent screen at a distance of 1.0 m. (No image intensifier is used in this example of an older imaging system.) Assume that the absorption efficiency of the intensifying screen is 100%, the conversion efficiency is 50%, the input exposure to the fluorescent screen is 1-R per second, the effective energy of the x-ray beam is 30 keV, and that $(\mu_{en}/\rho) = 0.1395$ cm^2/g. Also, assume an exposure time of 1 second.

(a) Compute the number of photons/cm² used to form the image at the input to the fluorescent screen.

(b) Compute the number of photons/cm² used to form the image at the output of the fluorescent screen.

(c) Compute the number of photons that enter the eye from a 1 cm² area of the fluorescent screen. Assume that the pupil of the eye has a diameter of 2 mm.

(d) Where is the quantum sink for this system?

(e) What is the system DQE?

Solution

(a) X-ray photons/cm² input at the screen:

The exposure is 1 Roentgen/s, so for a 1-second exposure time, the radiation exposure is 1 Roentgen.

At an effective energy of 30 keV for which

$$\left(\frac{\mu_{en}}{\rho}\right) = 0.1395 \text{ cm}^2/\text{g}, \tag{8.16}$$

the photon fluence is

$$\Phi_{\text{x-ray}} = \frac{0.00873\left[\frac{J}{\text{kg R}}\right]X}{E\left(\frac{\mu_{en}}{\rho}\right)_{\text{air}}}.$$

$$\Phi_{\text{x-ray}} = \frac{0.00873\left[\frac{J}{\text{kg R}}\right](1R)}{(30 \text{ keV/photon})\left(0.1395 \frac{\text{cm}^2}{\text{g}}\right)}\left[\frac{1 \text{ keV}}{1.6\times10^{-16} \text{ J}}\right]\left[\frac{1 \text{ kg}}{1000 \text{ g}}\right]$$

$$= 1.3\times10^{10} \frac{\text{photons}}{\text{cm}^2} \tag{8.17}$$

(b) Light photons/cm² output from the screen:

At a wavelength of $\lambda = 5000$ Å (Å = Angstroms), the energy of the light photon is

$$E = \frac{12,396 \text{ eV}\cdot\text{Å}}{5,000\text{Å}} \approx 2.5 \text{ eV}. \tag{8.18}$$

The total number of light photons generated by the screen having 50% conversion efficiency is

$$
\Phi_{\text{light}} = \frac{\left[1.3 \times 10^{10} \dfrac{\text{photons}}{\text{cm}^2}\right] 30 \times 10^3 \text{ eV} [0.5]}{2.5 \dfrac{\text{eV}}{\text{photon}}} = 7.8 \times 10^{13} \dfrac{\text{photons}}{\text{cm}^2}.
\tag{8.19}
$$

(c) Light photons entering the eye:

At a distance of 1 m, the number of photons from a 1 cm² area on the screen that enter the eye is

$$
\Phi_{\text{eye}} = 7.8 \times 10^{13} \text{ photons} \left[\frac{\pi(0.1 \text{ cm})^2}{4\pi(100 \text{ cm})^2}\right] = 1.95 \times 10^7 \text{ photons},
\tag{8.20}
$$

assuming isotropic emission by the screen and that $1/r^2$ drop off at 1 m.

(d) The quantum sink occurs at the stage where the image is formed by the fewest number of photons for the *same area of the object*.

This occurs within the eye of the radiologist.

(e) DQE of the system:

Assume that the input noise arises from the quantum statistics of the incident x-ray field and that the output signal noise is due to the number of light photons entering the eye (both assumed to be Poisson processes). Therefore, the DQE is

$$
\text{DQE} = \left(\frac{\text{SNR}_{\text{out}}}{\text{SNR}_{\text{in}}}\right)^2 = \frac{\Phi_{\text{eye}}}{\Phi_{\text{x-ray}}} = \frac{1.95 \times 10^7 \text{ photons}}{1.3 \times 10^{10} \text{ photons}} = 1.5 \times 10^{-3}.
\tag{8.21}
$$

Note: For a system consisting of a chain of detection component, SNR_{out} for the component representing the quantum sin c (i.e., the one with the lowest SNR) should be used for estimating the DQE.

8.6 NOISE SOURCES IN FILM-SCREEN SYSTEMS

Quantum mottle or statistical noise is only one of the several contributors to noise in an x-ray image. Both the film and the intensifying screen have a microstructure that contributes to small random fluctuations in the radiograph. In the film, the microstructure is due to the developed silver halide. The noise introduced by the silver grains is called the *film granularity*. Similarly, interactions of x-ray photons in the intensifying screen create flashes of light that expose the film. The random appearance of the light flashes reveals the underlying irregular structure of the screen called *structure mottle*. This irregular structure is due to an uneven distribution of phosphors in the screen, generating a spatially varying light distribution even when uniformly exposed to x-rays. Structure mottle occurs even when the x-ray exposure is infinite and quantum mottle is virtually eliminated. Structure mottle is spatially fixed, so rotating the screen rotates its pattern. Mathematically, we can define the three components of film-screen noise or mottle (σ^2) to be film granularity (σ_g^2), quantum mottle (σ_q^2), and screen structure mottle (σ_s^2). These are assumed to be independent processes, so these variances added to produce the overall screen-film noise variance.

8.6.1 Film Granularity

Nutting has shown that the density D of the film due to developed grains is related to the average number of developed grains per unit area (N_g/A) and the average developed grain area a_g according to the relationship:

$$D = (\log_{10} e)\left[a_g\left(\frac{N_g}{A} \right) \right].\tag{8.22}$$

Assuming that the only fluctuation in film density is due to the random number of film grains per unit area, and that other values are constant, the variance from film granularity σ_g is

$$\sigma_g^2 = \left(\frac{\partial D}{\partial N_g} \right)^2 \sigma_{N_g}^2 = \left[(\log_{10} e)\left(\frac{a_g}{A} \right) \right]^2 N_g \quad \text{and} \quad \sigma_g = \sqrt{(\log_{10} e)\left(\frac{a_g}{A} \right)D}.\tag{8.23}$$

8.7 NOISE IN ELECTRONIC IMAGING SYSTEMS

In electronic imaging systems such as those used in digital subtraction angiography (DSA), we have three principal sources of noise. The first arises from *quantum statistics*, in which the discrete nature of the radiographic signal (which often is photon-limited) introduces uncertainty into the image. The second is *electronic noise* that is generated in the detector or detector electronics. The third is due to *quantization error* that occurs in digital electronic imaging systems when the signal is digitized. These three components of noise will be presented separately, followed by a discussion on how they combine to contribute to the overall noise in the imaging system. In this section, we will focus on video-based image intensifier systems and resulting digital images; however, descriptions can be generalized to other electronic imaging systems used in diagnostic radiography.

8.7.1 Noise Introduced by Quantum Statistics

A fundamental noise source in digital x-ray images (as in most other medical images) is quantum statistical noise. If a radiographic signal is composed of N photons/pixel, then the uncertainty (i.e., standard deviation) in that signal is based on Poisson statistics.

For video-based image intensifier (II) systems, the uncertainty in the x-ray photon signal propagates into an uncertainty in the electronic signal, and this is important since the input to the II is the quantum sync for the II system. The amplitude of the resulting video noise can be calculated by assuming that the signal from the video camera is proportional to the x-ray photon fluence recorded by the image intensifier. This requirement is satisfied by a Plumbicon-type video camera. Assuming that the camera produces a maximum video voltage of V_{max} for N_{max} x-ray photons/pixel at the II input, then the video signal V for N photons/pixel leads to a proportionality relationship

$$\frac{V}{V_{max}} = \frac{N}{N_{max}} \Rightarrow V = \left(\frac{V_{max}}{N_{max}} \right) N, \tag{8.24}$$

where both N_{max} and V_{max} are constants based on the system configuration (Figure 8.6). The standard deviation of the video signal due to x-ray quantum statistical sources follows the same proportionality relationship:

$$\sigma_q = \left(\frac{V_{max}}{N_{max}} \right) \sigma_N, \tag{8.25}$$

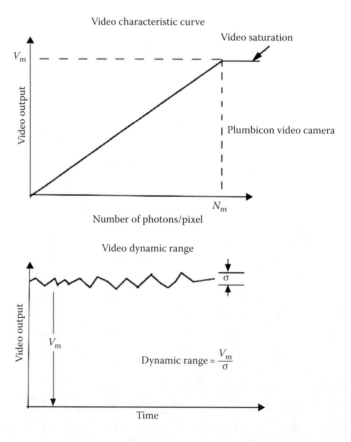

FIGURE 8.6 An ideal video camera produces an electronic signal proportional to the number of light photons/pixel. This relationship is maintained up to the maximum or saturation level of the video system (V_m) corresponding to N_m quanta per pixel at the input.

where the standard deviation σ_N is the uncertainty in x-ray photons/pixel

$$\sigma_N = \sqrt{N} \qquad (8.26)$$

so that the quantum noise seen in the video signal σ_q is

$$\sigma_q = \left(\frac{\sqrt{N}}{N_{max}} \right) V_{max}. \qquad (8.27)$$

Note: The units of this quantum noise are the same as measured for the video signal (volts).

8.7.2 Electronic Noise in a Video System

The electronic noise in a video camera is *independent of other sources of noise*. This electronic noise arises from the camera's dark current, that is, noise when no light is input to the camera. The electronic noise level is also assessed as a standard deviation (σ_e) that is related to V_{max} and the camera's dynamic range (D) as follows:

$$\sigma_e = \frac{V_{max}}{D}. \qquad (8.28)$$

The best vidicons have a dynamic range of ~1000:1 (although some are touted to achieve 2000:1 dynamic range). If the maximum video signal were 2 V, then the standard deviation of the electronic noise from a camera with a 1000:1 dynamic range would be 2 mV.

Unlike quantum statistical noise that varies as the square root of the number of quanta comprising the signal, the magnitude of electronic noise is not dependent on signal size. *It is therefore important to maximize the video signal whenever possible so that the electronic noise has a minimal contribution to overall noise.* Unfortunately, electronic noise becomes a problem in regions of image with low video levels that result from high object attenuation, especially if such areas are of diagnostic interest.

8.7.3 Noise due to Digital Quantization

Quantization noise is introduced into an analog signal when it is digitized. As with other forms of noise, we use variance to quantify the noise. Quantization variance is calculated as the variance between the quantized signal and the analog signal over a single quantization interval. The ADC's quantization interval Δ runs from $\mu_I - \Delta/2$ to $\mu_I + \Delta/2$, where the digitized signal is the fixed value μ_I. Quantization or digitization variance is calculated using Equation 8.29 that is greatly simplified by letting $\mu_I = 0$:

$$\sigma_\Delta^2 = \frac{1}{\Delta} \int_{\mu_I - \Delta/2}^{\mu_I + \Delta/2} (I - \mu_I)^2 \, dI = \frac{\Delta^2}{12}. \qquad (8.29)$$

Like electronic noise, the quantization noise does not relate to signal magnitude. Since we have some control over the magnitude of quantization noise, we can design imaging systems so that σ_Δ^2 is small in comparison to electronic and quantum noise sources. For example, if the variance due to electronic noise is σ_e^2, we can choose the quantization interval of the analog-to-digital converter (ADC) so that the variance of the quantization error is 1/10th the variance due to electronic noise. Doing this using Equation 8.29 yields

$$\sigma_\Delta^2 = \frac{\Delta^2}{12} = \frac{\sigma_e^2}{10}, \tag{8.30}$$

so the digitization interval is

$$\Delta = 1.1\sigma_e \approx \sigma_e. \tag{8.31}$$

Since this digitization interval (Δ) is approximately equal to σ_e, we can use this relationship to show that the digital dynamic range (D_{dig}) should be approximately equal to the analog electronic dynamic range (D).

$$\Delta \approx \sigma_e = \frac{V_m}{D} \approx \frac{V_m}{D_{dig}} \tag{8.32}$$

so $D_{dig} \approx D$.

The digital dynamic range is calculated as $D_{dig} = 2^N$ where N is the number of bits used by ADC. The ADC must provide a number of bits to cover the electronic dynamic range, that is, $2^N \geq D$. Using this approach, we see that for a video camera with a dynamic range of 1000:1, we would need at least a 10-bit ADC, that is, one with 1024:1 digital dynamic range. An ADC with one less bit would have a dynamic range of 512:1 and would not be adequate. An ADC with one more bit would have a dynamic range or 2048:1 and would be acceptable if other criteria (sampling rate, etc.) were sufficient for use with the video camera.

Example 8.4: DQE and Quantization Step Size

Assume that we digitize a signal with a quantization step Δ that is k times the standard deviation of the overall noise in the analog signal (σ). Calculate the DQE of the digital (output) versus analog (input) signal as a function of k. Discuss why your result makes sense physically.

Solution

The input signal to the digitizer has only analog noise, while the output signal has both analog and quantization noise. From Equation 8.29, we know that quantization noise variance for step size $\Delta = k\sigma$ is

$$\sigma_\Delta^2 = \frac{\Delta^2}{12} = \frac{(k\sigma)^2}{12}. \tag{8.33}$$

The total noise variance of the output digital signal is the sum of analog and quantization noise variances, since they are independent

$$\sigma_{total}^2 = \sigma^2 + \sigma_\Delta^2 = \sigma^2 + \frac{(k\sigma)^2}{12} \tag{8.34}$$

and assuming that the signal size S remains the same (i.e., the digitization introduces no gain), we know that the input and output SNRs are given by

$$SNR_{in} = S/\sigma \tag{8.35}$$

and

$$SNR_{out} = \frac{S}{\sigma_{total}} = \frac{S}{\sqrt{\sigma^2 + \frac{(k\sigma)^2}{12}}} = \frac{S/\sigma}{\sqrt{1 + \frac{k^2}{12}}} \tag{8.36}$$

so that the DQE is given by

$$DQE = \left(\frac{SNR_{out}}{SNR_{in}}\right)^2 = \frac{1}{1 + \frac{k^2}{12}}. \tag{8.37}$$

The DQE in this example is not an explicit function of analog noise level σ. As expected, for small k (where $\Delta \ll \sigma$), the DQE approaches unity so that the effect of digitization is minimal. For large k (where $\Delta \gg \sigma$), DQE tends toward zero because of the large contribution of digitization to system noise.

8.8 SYSTEM NOISE IN ELECTRONIC IMAGING SYSTEMS

As we saw in the aforementioned example, the system noise variance in an electronic imaging system is obtained by adding the variance from each noise component, assuming that these noise contributions are independent (or uncorrelated). *Also, the contribution to the signal and noise by digitization can be made sufficiently small so that it can be neglected.* The video camera signal V contains a time-varying component V_q proportional to the exposure to the input phosphor and a "random" time-varying term V_e arising from the "electronic noise" of the system (the time variable is implicit).

$$V = V_q + V_e. \tag{8.38}$$

The uncertainty in the video output can be calculated using propagation of errors

$$\sigma^2 = \sigma_q^2 + \sigma_e^2, \tag{8.39}$$

where σ, σ_q, and σ_e are the uncertainties in V, V_q, and V_e respectively.

We will investigate how these components contribute to system noise of a digital angiographic system in more detail in Chapter 10. We will complete this chapter with a brief description of how quantum statistical noise influences a digital subtraction angiographic system.

8.9 NOISE IN DIGITAL SUBTRACTION ANGIOGRAPHY (DSA)

In DSA, the image containing an artery after injection of an iodinated contrast agent is subtracted from the precontrast image. This procedure is performed to isolate the image of the opacified artery after the subtraction process removes the structure from background anatomy. We will show in Chapter 10 that the images should be subtracted only after they are logarithmically transformed. For now, we will assume that this is true and will use the propagation of error technique summarized in Chapter 7 to determine the contribution of quantum statistical noise in the digital subtraction image.

Assume that you have two images of a blood vessel obtained before and after iodine opacification, and you examine the number of photons/pixel within the bounds of the vessel. For the image obtained before opacification of the artery occurs, the number of photons/pixel is

$$N = N_0 e^{-\mu_w x}, \tag{8.40}$$

where

μ_w is the linear attenuation coefficient of water (i.e., blood and soft tissue)
x is the thickness of the body
N_0 is the incident number of photons/pixel

After opacification, the attenuation increases due to the iodinated contrast agent in the artery so that

$$N_I = N_0 e^{-(\mu_w x + \mu_I t)}, \tag{8.41}$$

where

μ_I is the linear attenuation coefficient of the iodinated contrast agent
t is the thickness of the artery

We will assume that attenuation by the blood in the artery has not changed. The signal in the logarithmic subtraction image is formed as

$$S = \ln(N) - \ln(N_I) = \mu_I t. \tag{8.42}$$

Propagation of errors for Equation 8.42 estimates the variance in the signal S as

$$\sigma^2 = \left[\frac{\partial S}{\partial N} \right]^2 \sigma_N^2 + \left[\frac{\partial S}{\partial N_I} \right]^2 \sigma_{N_I}^2. \tag{8.43}$$

Since the photon fluence behaves according to Poisson statistics, we have

$$\sigma_N^2 = N \tag{8.44}$$

and

$$\sigma_{N_I}^2 = N_I \tag{8.45}$$

so that from Equations 8.43 to 8.45, we have

$$\sigma^2 = \left(\frac{1}{N^2}\right)N + \left(\frac{1}{N_I^2}\right)N_I = \frac{1}{N} + \frac{1}{N_I} = \frac{e^{\mu_w x}\left(1 + e^{\mu_I t}\right)}{N_0}. \tag{8.46}$$

For small values of $\mu_I t$ (i.e., when $\mu_I t \ll 1$),

$$e^{\mu_I t} = 1 + \mu_I t \tag{8.47}$$

and

$$\sigma^2 = \frac{e^{\mu_w x}\left[2 + \mu_I t\right]}{N_0} \tag{8.48}$$

so that the SNR of DSA (ignoring electronic noise) is

$$\text{SNR} = \frac{\mu_I t}{\sqrt{e^{\mu_w x}\left[2 + \mu_I t\right]}}\sqrt{N_0}. \tag{8.49}$$

HOMEWORK PROBLEMS

P8.1 A radiologist views the output phosphor of an image intensifier at a distance of 1 m. Assume that the x-ray absorption efficiency is 75%, the conversion efficiency of the input phosphor and output phosphor are both 50%, that the input exposure to the fluorescent screen is 10^{-5} R/s, and that the brightness gain (minification × flux gain) of the image intensifier is 10,000 (May need to review principles of IIs). Assume that the pupil of the eye has a diameter of 2 mm when the output screen is viewed and that the effective energy of the x-ray beam is 30 keV and that at this energy

$$\left(\frac{\mu_{en}}{\rho}\right)_{air} = 0.1395 \text{ cm}^2/\text{g}.$$

Compute the number of photons used to form the image for the following:

(a) X-rays absorbed within the input phosphor.

(b) Light image emitted at the surface of the output phosphor.

(c) The light field entering the eye.

(d) Where is the quantum sink for this system?

(e) Compare this result with the one derived in the notes in which the observer views a conventional fluorescent screen directly without the benefit of an image intensifier.

P8.2 For a thesis project, a student wants to make a bone densitometer that measures the thickness of bone and soft tissue with a dual-energy detection system. The measurement of bone depends on two independent measurements with different energy radioisotopes.

(a) If I_{01} is the incident photon fluence in the first measurement (energy 1) and I_{02} is the incident photon fluence with the second measurement (energy 2), show that the thickness b of bone is equal to

$$b = \frac{\mu_{t2} \ln\left(\dfrac{I_{01}}{I_1}\right) - \mu_{t1} \ln\left(\dfrac{I_{02}}{I_2}\right)}{\mu_{t2}\mu_{b1} - \mu_{t1}\mu_{b2}},$$

where μ_{t1} and μ_{t2} are the linear attenuation coefficients of tissue at energies 1 and 2, respectively, and μ_{b1} and μ_{b2} are the linear attenuation coefficients of bone at energies 1 and 2.

(b) Use propagation of errors to show that the variance in the measurement of the bone thickness b is

$$\sigma_b^2 = \left(\frac{\mu_{t2}}{\mu_{t2}\mu_{b1} - \mu_{t1}\mu_{b2}}\right)^2 \left(\frac{1}{I_{01}} + \frac{1}{I_1}\right) + \left(\frac{\mu_{t1}}{\mu_{t2}\mu_{b1} - \mu_{t1}\mu_{b2}}\right)^2 \left(\frac{1}{I_{02}} + \frac{1}{I_2}\right) = \frac{A_1}{I_{01}} + \frac{A_2}{I_{02}},$$

where

$$A_1 = \left(\frac{\mu_{t2}}{\mu_{t2}\mu_{b1} - \mu_{t1}\mu_{b2}}\right)^2 \left[1 + e^{(\mu_{t1}t + \mu_{b1}b)}\right]$$

and

$$A_2 = \left(\frac{\mu_{t1}}{\mu_{t2}\mu_{b1} - \mu_{t1}\mu_{b2}}\right)^2 \left[1 + e^{(\mu_{t2}t + \mu_{b2}b)}\right].$$

Note that for a given patient, A_1 and A_2 are both constant.

(c) In a real imaging system, we want to maximize the precision of the measurement (i.e., minimize σ_b^2) for a certain entrance dose to the patient. If f_1 and f_2 are the ratios of the radiation absorbed dose to photon fluence at energies 1 and 2, respectively, and if a fraction k of the dose is obtained from photons of energy 1, while the remaining fraction $(1 - k)$ of the dose is obtained from photons of energy 2, and if $D_0 = D_1 + D_2$ is the total dose received by the patient, then show that

$$\sigma_b^2 = \frac{f_1 A_1}{k D_0} + \frac{f_2 A_2}{(1-k) D_0}.$$

(d) Differentiate this equation with respect to k to determine what fractional mix of photons at energies 1 and 2 will give the best precision for the given dose D_0. Remember that $f_1, f_2, A_1, A_2,$ and D_0 are all constant. Show that the best precision is obtained when the dose delivered by the photons of energy 1 contributes a fraction k_{min} of the total dose where

$$k_{min} = \frac{1}{\sqrt{\frac{f_2 A_2}{f_1 A_1}} + 1}.$$

(e) Finally, show that this condition is satisfied when the source intensities I_{01} and I_{02} are related by the formula

$$\frac{I_{01}}{I_{02}} = \frac{k_{min} f_2}{(1 - k_{min}) f_1}.$$

P8.3 Calculate the DQE for a system in which

(a) The detector noise variance is three times larger than quantum noise variance. Assume 100% x-ray detection efficiency.

(b) The system noise variance is twice the quantum noise variance and for which one-half of the incident x-rays are detected.

P8.4 You are developing an image processor to digitize a video signal from an image intensifier system.

(a) Derive the digital level spacing Δ so that the noise contributed by digitization is approximately 30% of the analog noise σ_{analog}. Express your answer in terms of σ_{analog}.

(b) If the dynamic range of the television camera is 1000:1 and the peak output of the television camera is 2 V, calculate the digitization spacing (in terms of millivolts) and the number of bits needed to digitize the signal.

(c) If you know that the quantum noise always will be at least as large as the electronic noise of the system, how does this affect your answer to (b).

P8.5 Using the equations derived in class, compare the SNR (at peak video signal) associated with a TV fluoroscopy system with a dynamic range of 1000:1 for the following cases. Assume that the dynamic range refers to a characteristic resolution element of (1 mm^2).

(a) Fluoroscopy at an exposure of 1 R per image.

(b) Digital radiography at an exposure of 1 R per image (neglect digitization noise).

III

Advanced Concepts

CHAPTER 9

Noise-Resolution Model

9.1 DESCRIPTION AND DEFINITION

The autocorrelation function is a measure of similarity between a data series and a shifted copy of the series as a function of shift magnitude. It is based on correlation analysis, which is used to find periodic patterns in noisy data, characterize similarity patterns for data compression, and measurement of spatial resolution of an image receptor with uniform white noise as the input. For medical imaging, a classic use of the autocorrelation function is in the measurement of film and screen spatial resolution, more generally detector resolution.

The autocorrelation function is defined as $E\{I(x)\ I(x + \Delta)\}$, where $E\{\}$ is the expectation operation and $I(x)$ is a 1-D function and Δ the shift in x. A similar function, the autocovariance function, is defined as $E\{[I(x) - <I(x)>][I(x + \Delta) - <I(x)>]\}$, and if $I(x)$ is a zero-mean function, then $<I(x)> = 0$ and the two definitions give the same result. In fact, for most medical imaging applications, $I(x)$ would be transformed to a zero-mean function before the autocorrelation function is calculated, so for our purpose, autocorrelation function = autocovariance function.

The autocorrelation function will be designated using a capital letter as $C_x(\Delta)$ as follows:

$$C_x(\Delta) = E\{I(x)I(x + \Delta)\}. \tag{9.1}$$

When probability density functions are not known (very likely), the expected value is calculated as the average.

Example 9.1: Autocorrelation Function of Simple Periodic Function

The following example shows how to calculate the autocorrelation function for a simple series of data that repeats every three samples. First let us take a look at $C_x(\Delta)$ with $\Delta = 0$:

x	1	2	3	4	5	6	7	8	9
$I(x)$	... +1,	−1,	0,	+1,	−1,	0,	+1,	−1,	0, ...
$I(x + 0)$	... +1,	−1,	0,	+1,	−1,	0,	+1,	−1,	0, ...
$I(x)I(x + 0)$	... +1,	+1,	0,	+1,	+1,	0,	+1,	+1,	0, ...

$$C_x(0) = \frac{1}{N}\sum_{x=1}^{N} I(x)I(x+0) = \frac{1}{9}\sum_{x=1}^{9} I(x)I(x+0) = \frac{6}{9} = \frac{2}{3} \qquad (9.2)$$

For zero shift, the $C_x(\Delta)$ in this example is just the average of $I(x)*I(x)$ over each period. The average was taken over three periods ($N = 9$) in this example, but in general, the extent of x would be much larger. Also, note that the numerator of C increases by 2 for each period of the sample, while the denominator increases by 3 such that $C_x(0) = 2/3$ when averaged over exact multiples of periods. For a large number of samples (i.e., for $N = 1000$), the exact number of periods is not as important. For example, since 999 points of the 1000 would cover 333 complete periods from this example, only 1 sample would not contribute to the sum. If the period is known, then the calculation of the autocorrelation function should span multiples of the period.

Now calculate $C_x(\Delta)$ with $\Delta = 1$

x	1	2	3	4	5	6	7	8	9
$I(x)$	... +1,	−1,	0,	+1,	−1,	0,	+1,	−1,	0, ...
$I(x + 1)$	... −1,	0,	+1,	−1,	0,	+1,	−1,	0,	+1, ...
$I(x)I(x + 1)$	... −1,	0,	0,	−1,	0,	0,	−1,	0,	0, ...

$$C_x(1) = \frac{1}{N}\sum_{x=1}^{N} I(x)I(x+1) = \frac{1}{9}\sum_{x=1}^{9} I(x)I(x+1) = \frac{-3}{9} = \frac{-1}{3} \qquad (9.3)$$

Note that as the numerator in this calculation changes by −1 for each period, the denominator changes by +3, therefore $C_x(1) = −1/3$. If the calculations are continued for various spacings, you will get the following:

$$\Delta = \dots -3, -2, -1, 0, +1, +2, +3 \dots$$

$$C_x(\Delta) = \dots 2/3, -1/3, -1/3, 2/3, -1/3, -1/3, 2/3 \dots \qquad (9.4)$$

Inspection of this autocorrelation function leads to the following characteristics:

- The autocorrelation function is positive (here = 2/3) at $\Delta = 0$. Further, it can be shown that the value of $C_x(0) \geq C_x(\Delta)$ for all Δ. For nonrepeating functions $C_x(0) > C_x(\Delta)$, that is, $C_x(\Delta)$ will be a maximum at $\Delta = 0$. This will clearly be the case when we investigate detector spatial resolution using random noise.
- The autocorrelation function repeats with a period = 3, the same period as $I(x)$.
- The autocorrelation function is −1/3 for $\Delta = \pm 1, \pm 2$. The autocorrelation function is identical for these two displacements, that is, symmetric about $\Delta = 0$.

Autocorrelation (or autocovariance) with $\Delta = 0$ is just the variance of $I(x)$ if it is a zero-mean function, or stated mathematically $C_x(0) = \sigma^2$. With this identity in mind, we can interpret $C_x(\Delta)$ as the covariation (or correlation) between a function $I(x)$ and its shifted version $I(x + \Delta)$. A normalized measure of autocorrelation, normalized relative to the variance in $I(x)$ at $\Delta = 0$ (its maximum), is $R_x(\Delta) = C_x(\Delta)/C_x(0)$. Therefore, $R_x(0) = 1$ and $R_x(\Delta) \leq 1$ for all other values of Δ, providing a standard measure of fractional similarity as a function of displacement.

Example 9.2: Autocorrelation Function of Sinusoids

Since cosine is a periodic function, we must integrate over an integer number of periods to ensure appropriate calculation of its autocorrelation function. The basic equation is

$$C(\Delta) = \frac{1}{nT} \int_{\frac{-nT}{2}}^{\frac{nT}{2}} \cos\left(\frac{2\pi}{T} x\right) \cos\left[\left(\frac{2\pi}{T}\right)(x + \Delta)\right] dx. \tag{9.5}$$

From trigonometry $\cos A \cos B = 1/2 \cos(A - B) + 1/2 \cos(A + B)$ and setting

$$A = (2\pi/T)x \quad \text{and} \quad B = (2\pi/T)x + (2\pi/T)\Delta \text{ and using } \cos(\theta) = \cos(-\theta),$$

we get

$$C(\Delta) = \frac{1}{nT} \int_{\frac{-nT}{2}}^{\frac{nT}{2}} \frac{1}{2} \cos\left(\frac{2\pi}{T} \Delta\right) dx + \frac{1}{nT} \int_{\frac{-nT}{2}}^{\frac{nT}{2}} \frac{1}{2} \cos\left[\left(\frac{2\pi}{T}\right)(2x + \Delta)\right] dx$$

$$C(\Delta) = \left(\frac{1}{2} \cos\left(\frac{2\pi}{T} \Delta\right)\right) \left(\frac{1}{nT} \int_{\frac{-nT}{2}}^{\frac{nT}{2}} dx\right) \tag{9.6}$$

$$C(\Delta) = \frac{1}{2} \cos\left(\frac{2\pi}{T} \Delta\right), \tag{9.7}$$

since the second integral is zero when integration range is a multiple of the period (T).

An identical answer is seen for the sine function. Note that $C(\Delta) = 1/2$ for $\Delta = 0$ and $C(\Delta)$ has the same period (T) as the sinusoid.

In general, the following are true for all autocorrelation functions for *zero-mean* $I(x)$:

- $C(0) = \sigma^2$ at $\Delta = 0$, the autocorrelation function is variance of $I(x)$.
- $C(-\Delta) = C(\Delta)$, autocorrelation functions are symmetric.
- $C(0) \geq C(\Delta)$, maximum at zero displacement ($\Delta = 0$).
- $C(\Delta) = \sigma^2(\Delta)$, autocorrelation function = variance as function of Δ.
- $R(\Delta) = C(\Delta)/C(0)$, normalized autocorrelation function.

9.2 AUTOCORRELATION AND AUTOCONVOLUTION FUNCTIONS

There is a natural similarity between the autocorrelation function and convolving a function with itself (autoconvolution). Autoconvolution is as follows:

$$f \otimes f = \int f(y)f(x-y)dy, \tag{9.8}$$

whereas for autocorrelation, there is *no reflecting*, only shifting, in the second term leading to

$$C(x) = \int f(y)f(y-x)dy = \int f(y)f(y+x)dy, \tag{9.9}$$

where

 y is a dummy x variable for integration

 $f(x)$ is a real (i.e., not complex) function

The x in these equations can be replaced by Δ, since both indicate a shift in y. Graphing the two forms in Equation 9.9 will reveal why they are identical on integration. Also, intuitively, the shift direction can be to the right or left as long as all possible combinations are used.

9.3 POWER SPECTRAL DENSITY FUNCTION

The Fourier transform of $C(\Delta)$ has a simple form as follows:

$$\varphi(u) = \Im\{C(\Delta)\} = F(u)F^{\star}(u) = \left|F(u)\right|^{2}. \tag{9.10}$$

$\varphi(u)$ is called the "power spectral density" (PSD) of $f(x)$. *Equation 9.10 can be used to directly calculate the autocorrelation function using* $C(\Delta) = \Im^{-1}\left\{\left|F(u)\right|^{2}\right\}.$

The PSD and the autocorrelation function are Fourier transform pairs, with the following form in 2-D:

$$\varphi(u,v) = \iint C(\Delta_x, \Delta_y)e^{-2\pi i(u\Delta_x + v\Delta_y)}d\Delta_x d\Delta_y \tag{9.11}$$

and

$$C(\Delta_x, \Delta_y) = \iint \varphi(u,v)e^{2\pi i(u\Delta_x + v\Delta_y)}dudv. \tag{9.12}$$

In general, the following are true for the PSD function:

- $C(0,0) = \sigma^2 = \iint \varphi(u,v)dudv$ Integral of PSD = variance $f(x)$.

- $\varphi(u, v)$ PSD is real (Equation 9.10).

TABLE 9.1   General Form of Autocorrelation and PSD Functions

Input Function	$C(\Delta)$	$\phi(u)$
Sinusoid ($\mu = 0$)	Cosine	Delta functions
Gaussian	Gaussian	Gaussian
Delta function	Delta function	Constant
Random noise ($\mu = 0$)	Delta function	Constant

- $\varphi(u, v) \geq 0$ PSD is nonnegative (Equation 9.10).
- $\varphi(u, v) = \varphi(-u, -v)$ PSD is symmetric.

A summary of autocorrelation and PSD functions for various input functions (Table 9.1) is helpful in understanding their use.

9.4 WIENER SPECTRUM

The PSD of zero-mean random noise is called the *Wiener spectrum* and is usually written as $W(u)$ rather than $\varphi(u)$ to make this explicit. The autocorrelation function $C(\Delta)$ and Wiener spectrum $W(u)$ properties of random noise can be exploited to determine the spatial resolution of film/screen systems. Figure 9.1 is a 256×256 random noise image $i(x, y)$ with mean = 0 and standard deviation = 1. Figure 9.2 is its Wiener spectrum $\{|I(u, v)|^2\}$ where $|I(u, v)|$ is the magnitude of the Fourier transform of $i(x, y)$. The origin $u = 0$, $v = 0$ is at the center of Figure 9.2.

FIGURE 9.1   2-D image $[i(x, y)]$ of random values ($m = 0$, $\sigma = 1$).

FIGURE 9.2 $|I(\mu, v)|^2$ is the power or Wiener spectrum of $i(x, y)$.

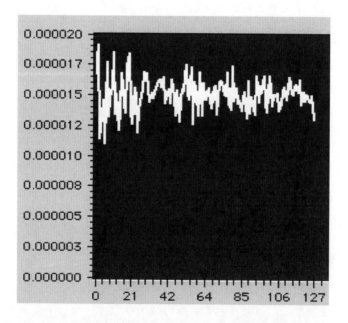

FIGURE 9.3 Radial frequency plot $W(r)$ of Wiener spectrum from Figure 9.2.

The integral of $|I(u, v)|^2$ is equal to 1 as predicted, since it should be equal to $C(0,0) = \sigma^2 = 1$ for the image $i(x, y)$. Figure 9.3 is a graph of the Wiener spectrum $[W(\rho)]$ expressed as a function of the distance from the origin ρ (i.e., the radial frequency). Note that the highest frequency is ~128 cycles or line pairs as expected for a 256×256 image. Note also that for this graph, the data appear to vary about a mean response. Since every location in $W(u, v)$ should vary randomly about its mean and the sum of all locations is equal to 1, then the mean value can be estimated as sum/256² = 1/256² =1.53 × 10⁻⁵.

If the Wiener spectrum in Figure 9.3 were fitted with a straight line, the slope would be ~0 with intercept equal to the mean value (1.53 × 10⁻⁵). The constant magnitude (other than noise) across all possible frequencies demonstrates that the Weiner spectrum of random noise, if unmodified by the system transfer function, is made up of equal amplitudes at all frequencies. This noise is often called "white" due to the fact that white light also has a uniform mix of light of over a broad range of frequencies or wavelengths. Figures 9.3 is an example of the Wiener spectrum for an ideal imaging system, that is, one with a constant magnitude system frequency response $|H(\rho)| = k$. *For a real imaging system, the Wiener spectrum $W(\rho) = |H(\rho)|^2$ is the square of the normalized spatial frequency response.*

This relationship can therefore be used to test the resolution capabilities of an imaging system when the input is "white" noise.

Figures 9.4 through 9.6 provide insight into how the Wiener spectrum relates to system resolution. Let us assume that an imaging system alters the 2-D random image $i(x, y)$ of Figure 9.1 due to blurring as a result of point spread function. Figure 9.4 models this using a 9 × 9 Gaussian PSF applied to $i(x, y)$ to simulate the imaging system blurring. The blurring is modeled in the frequency domain as $I_s(u, v) = I(u, v)H(u, v)$, where $I(u, v)$ is the Fourier transform of $i(x, y)$ and $H(u, v)$ is the simulated Gaussian system transfer function.

The 2-D Wiener spectrum of the blurred noise image from Figure 9.4 is given in Figure 9.5 and as a radial plot $[W_s(\rho)]$ in Figure 9.6. It follows that $W_s(\rho) = |H_s(\rho)|^2$ or $|H_s(\rho)| = [W_s(\rho)]^{1/2}$. This latter equation states that the square root of the Wiener spectrum for a system is the magnitude of its frequency response. Therefore, $|H_s(\rho)|$ is similar to MTF. *The MTF is estimated as $[W_s(\rho)/W_s(0)]^{1/2}$, where the division ensures that MTF(0) = 1.*

FIGURE 9.4 Random noise image of system modeled with a 9 × 9 Gaussian point spread function.

FIGURE 9.5   Wiener spectra of system modeled with a 9 × 9 Gaussian point spread function.

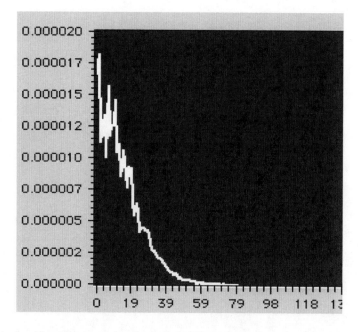

FIGURE 9.6   Magnitude of frequency spectrum $F(r)$ for Figure 9.5.

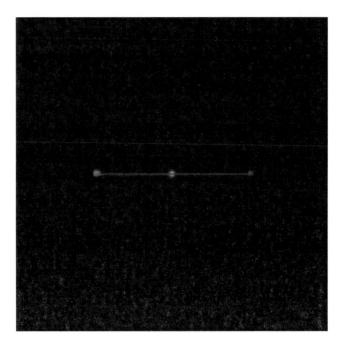

FIGURE 9.7 Autocorrelation of zero-mean random noise $i(x, y)$ smoothed with a 9 × 9 Gauss-ian filter.

While the Wiener spectrum can be calculated without using the autocorrelation function, it is instructive to analyze system's spatial blurring by direct analysis of the autocorrelation function of noise. The rationale is that the autocorrelation function for random noise input to an ideal imaging system simulates a perfect point spread function for testing, that is, expected response $C_r(\Delta x, \Delta y) = \delta(x, y)$. Broadening of this expected point response is due to imaging system blurring, that is, the PSF(x, y) of the system. We say that the system PSF introduces short-range correlations, and this is reflected as a broadening of the auto-correlation function of random noise. In the spatial domain, this is modeled as

$$C_s(\Delta x, \Delta y) = C_r(\Delta x, \Delta y) \otimes [\text{PSF}(-x, -y) \otimes \text{PSF}(x, y)], \qquad (9.13)$$

where the broadened autocorrelation function $C_s(\Delta x, \Delta y)$ is different from the ideal auto-correlation function due to convolution with PSF(x, y) of each of the image functions in $C_r(\Delta x, \Delta y) = \delta(x, y)$. An example of this is provided in Figures 9.7 and 9.8.

The similarity of the autocorrelation analysis of a random noise image and the direct application of smoothing is seen from the similarity of Figures 9.7 through 9.10. The square root of the profile curves in Figures 9.8 and 9.10 provides an estimate of the system PSFs.

9.5 MEASURING FILM/SCREEN RESOLUTION

The autocorrelation function and Wiener spectra are both sensitive to low-frequency back-ground variations and this limits their use in nuclear medicine, where such changes in uniformity are often present. However, both can be used to assess spatial resolution of

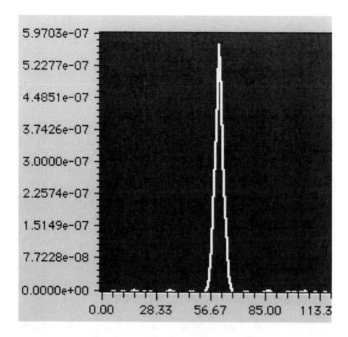

FIGURE 9.8 Profile along the autocorrelation function $C(\Delta x, \Delta y)$ in Figure 9.7.

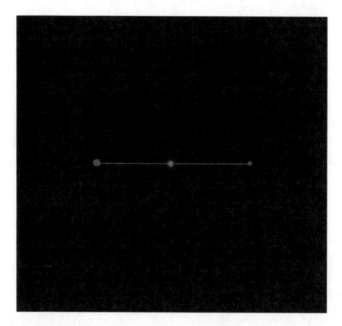

FIGURE 9.9 Point smoothed twice with a 9×9 Gaussian filter.

film/screen systems without the confounding effects of other components of spatial resolution. A uniform random image can be acquired by exposing a small region of a film/screen at a large distance from the x-ray tube. *This image is free of effects of focal spot size, grid, or magnification.* The film is then scanned with a small aperture microdensitometer and the recorded film density converted to relative exposure using calibration data for the

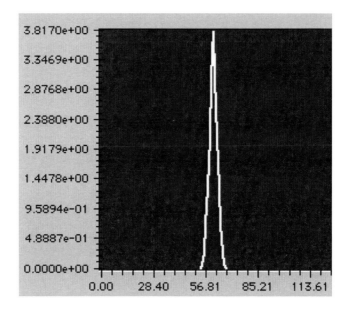

FIGURE 9.10   Profile along smoothed point in Figure 9.9.

film/screen combination. The film/screen MTF is calculated as the square root of the normalized $W(\rho)$ for these data. If necessary, the MTF can be corrected for the drop-off due to the scanner aperture, but with a sufficiently small aperture, this is not needed (one with drop in frequency response <5% at the highest frequency of interest).

The uniform noise presented to the film/screen system is quantum noise (i.e., white noise spectrally). The added noise in the developed image is also due to random processes in both the screen and the film, and since these are assumed to be independent, the random noises for these are additive. However, blurring by film and screen alter the MTF, which leads to an overall MTF with decreasing response as frequency increases (Figure 9.11).

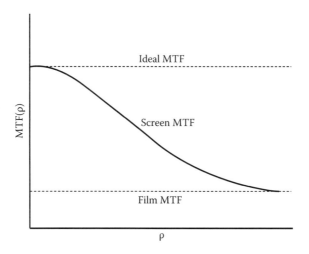

FIGURE 9.11   Example MTF for film/screen system.

Since the film MTF is approximately constant over the range when the screen MTF is falling off, these can be separated.

9.6 MORE ON DQE

The system equation for DQE given in Chapter 8 does not account for the natural loss in signal contrast with increasing frequency of imaging systems (Equation 8.2). Since mean random noise levels at the input and output of an imaging system are approximately constant with increasing frequency, the change in signal contrast or modulation with frequency leads to changes in SNR(f). A more general equation for DQE is therefore

$$DQE(f) = \frac{SNR_{out}^2(f)}{SNR_{in}^2(f)}.$$

(9.14)

Using logic that was applied for the development of the Rose model equation, $SNR_{in}(f)$ can be expressed as a function of input contrast $c_{in}(f)$ and N_{in}, the number of photons from an area of interest within the background:

$$SNR_{in}^2(f) = c_{in}^2(f)N_{in} = MTF_{in}^2(f)N_{in},$$

(9.15)

where for sinusoids contrast = modulation such that $c_{in}^2(f) = MTF_{in}^2(f)$. Since we can assume an ideal frequency response at the input of the system, $MTF_{in}^2(f) = 1$ for all frequencies, simplifying SNR_{in}:

$$SNR_{in}^2(f) = MTF_{in}^2(f)N_{in} = 1 \cdot N_{in} = N_{in}.$$

(9.16)

We can calculate $SNR_{out}(f)$ for a sinusoid using $signal_{out}(f) = c_{out}(f)$ times the mean signal out as follows:

$$SNR_{out}^2(f) = \frac{c_{out}^2(f)\bar{s}_{out}^2}{\sigma_{out}^2} = MTF_{out}^2(f)\left(\frac{\bar{s}}{\sigma}\right)_{out}^2.$$

(9.17)

Again for sinusoidal response, $MTF_{out}^2(f)$ is equivalent to $c_{out}^2(f)$. $MTF_{out}(f)$ is determined by the overall imaging system frequency response, that is, due to focal spot size, magnification, grids, scatter, film, screen, etc. The average signal out ($\bar{s}_{out}$) and standard deviation (σ_{out}) are calculated using the system's output units. The ratio of Equation 9.17 to Equation 9.16 gives the system DQE(f):

$$DQE(f) = \frac{(\bar{s}/\sigma)_{out}^2}{N_{in}} MTF_{out}^2(f),$$

(9.18)

where

$$DQE(0) = \frac{\left(\bar{s}/\sigma\right)^2_{out}}{N_{in}}, \tag{9.19}$$

since $MTF^2_{out}(0) = 1$. Note that the mean and standard deviation in Equation 9.19 are just those that would be measured if the signal were uniform, and DQE(0) is the same as the DQE we determined using Equation 8.2.

A simpler version for $DQE(f)$ is as follows:

$$DQE(f) = DQE(0) \cdot MTF^2_{out}(f). \tag{9.20}$$

Since $MTF^2_{out}(f) \leq 1$, *the DQE(f) drops off with increasing frequency indicating less effective use of the input quanta.*

$DQE(f)$ for subcomponents of a system can be calculated using appropriately determined $SNR_{in}(f)$ and $SNR_{out}(f)$ for each component, a case where $SNR_{in}(f)$ is no longer a constant.

In most applications, $MTF(f)$ is calculated from the point spread function. It may be necessary to correct for sample spacing and aperture size used when digitizing the output image to obtain an accurate value for $MTF_{out}(f)$, that is, if these are not sufficiently small.

HOMEWORK PROBLEMS

P9.1 In Chapter 5, we found that the convolution operation in the spatial domain becomes multiplication in the frequency domain. In the autocorrelation chapter, we noted a similar property for autocorrelation. However, we never looked at the finer details of this complex multiplication in the frequency domain.

Given $F(u) = F_r(u) + iF_i(u)$ and $G(u) = G_r(u) + iG_i(u)$ where the subscripts r and i indicate the real and imaginary parts of these complex functions. Determine the following:

(a) The full expression (both real and imaginary parts) for $F(u) \cdot G(u)$ in the frequency domain. This relates to how we use the FFT to calculate convolution.

(b) The full expression for $\bar{F}(u) \cdot F(u)$ in the frequency domain. This relates to how we use the FFT to calculate autocorrelation. *The bar over F(u) indicated its complex conjugate.*

(c) The full expression for $\bar{F}(u) \cdot G(u)$ in the frequency domain. This relates to how we use the FFT to calculate cross-correlation.

P9.2 Using Excel, Mathcad, or any similar program, create a 256-point zero-mean random noise sample $i(x)$.

(a) Calculate mean and variance of $i(x)$.

(b) Calculate autocorrelation function $C(\Delta)$ for $i(x)$. Describe how you did this and any assumptions you made. What is the relationship between $C(0)$ and $i(x)$?

(c) Calculate Wiener spectrum $W(u)$ for $i(x)$. What is the relationship between $C(0)$ and $W(u)$? How would you convert $W(u)$ to $\text{MTF}(u)$?

P9.3 Three images are provided as follows: (i) X_bar.nii.gz, (ii) Ring.nii.gz, and (iii) 3-bars. nii.gz. Use *Mango* to calculate the autocorrelation image for each and verify that the peak value in the autocorrelation image is equal to the sum in the original image.

(a) In the X_bar autocorrelation image, illustrate using the cross-section graph a line oriented at 0°, 45°, and 90°. Make sure that the line you use goes through the peak value of the autocorrelation image. Why is the symmetry different in the horizontal and vertical directions?

(b) For the ring autocorrelation image, what is the diameter of the middle of the outer ring? A line through the peaks in the graph does not go to zero between the outer ring and the bright central section. Why?

(c) For the 3_bars autocorrelation image, why are there five horizontal bands? The cross-section graph for a vertically oriented line through the autocorrelation peak goes to zero between the peaks. Why? Also why are the peak values in this curve different? Do they conform to what should be expected?

CHAPTER 10

The Rose Model

10.1 INTRODUCTION

The model of human visual perception introduced in Chapter 1 was named after its formulator Albert Rose. Dr. Rose was a scientist at the Radio Corporation of America (RCA) investigating the basic operating parameters of television in the 1940s and 1950s. In particular, he was trying to relate levels of contrast, resolution, and noise.

Isaac Newton once said, "If I have seen further, it is by standing on the shoulders of giants." This has occurred over and over in science, for example, when Johannes Kepler built his model of the heliocentric universe based on the painstaking data of planetary motion collected by Tycho Brahe. In much the same way, Albert Rose built his model of human visual perception on the painstaking data collected by Richard Blackwell. We therefore will digress and talk about Blackwell's studies, before continuing with the discussion of the Rose model and its application in diagnostic radiology.

Richard Blackwell was a scientist who worked on visual perception studies for the United States Navy during World War II. The navy was interested in what level of light and how large of an object was required by a sailor to spot an enemy vessel at night. It is obvious that a large light is easier to see than a smaller light, and that a bright light is easier to see than a dim light. But is a large dim light easier to see than a small bright light? The navy (or someone in the navy) wanted to know the answer to this question and provided Blackwell with funds to conduct this research.

The by-gone days of governmental generosity for research are apparent in the study that Blackwell performed. For his work, he hired 20 young women and kept them housed and fed in a dormitory built close to his laboratory. For 2 years, he had the women observe simple images of gray circles (the targets) on plain backgrounds projected onto a screen. For each observation, each woman reported whether or not she saw the circle, and in which quadrant of the projection screen it was located. Blackwell and his young female subjects performed thousands of observation studies, and slowly out of this painstaking work emerged a pattern that related target size and contrast to the level of illumination (or noise level) at the threshold of visualization. The results published in graphical form by Blackwell formed the basis for the more theoretical work by Rose.

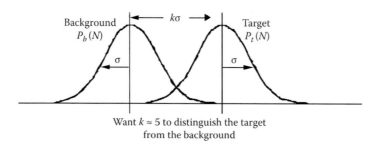

Background area of same size imaged with N_b photons

Target imaged with N_t photons

Both N_b and N_t will be randomly (Poisson) distributed with a standard deviation of σ

Background $P_b(N)$ $k\sigma$ Target $P_t(N)$

σ σ

Want $k \approx 5$ to distinguish the target from the background

FIGURE 10.1 Probability distributions for background and target at the threshold of detection.

The theory, outlined by Rose, basically is a probabilistic model of low-contrast threshold detection (Figure 10.1). The Rose model states that an observer can differentiate two regions of the image, called "target" and "background," only if there is sufficient information to do so. Specifically, the "signal" (defined as the difference in the mean number of photons in each region) must be k times the "noise" or $k\sigma$ where the noise is the statistical uncertainty (standard deviation) σ in the regions. Based on the data of Blackwell, Rose found that k in the range of 5–7 was the threshold for visual detection of a target relative to its background. *As defined "k" is a signal-to-noise ratio (SNR = signal/noise = $k\sigma/\sigma = k$), and therefore, "k" specifies the minimum SNR required for visual detection.*

10.2 DERIVATION OF THE ROSE MODEL

We will derive the Rose model equation using a simple statistical model, which assumes that the number of photons in the target and background is Poisson distributed. Furthermore, we will apply this concept to low-contrast situations where the object of interest (target) is just visually perceptible (i.e., where the target and background numbers are nearly identical).

Given:
  N is the number of photons in the background area
  ΔN is the difference in number of photons between target and background areas
  A is the area of the target (same area used for background)
  C is the contrast of the signal with respect to the background

The contrast is defined as follows:

$$C = \Delta N/N. \tag{10.1}$$

This difference ΔN is the signal of interest, and rearranging Equation 10.1 leads to an expression for the signal in terms of the contrast and the number of photons:

$$\text{Signal} = \Delta N = C \cdot N. \tag{10.2}$$

At the threshold of visualization, ΔN is small, so the number of photons in target and background areas is approximately equal. The noise is then the square root of N:

$$\text{Noise} = \sqrt{N}. \tag{10.3}$$

The SNR (k) at the threshold of visualization is then

$$k = \frac{\text{Signal}}{\text{Noise}} = \frac{C \cdot N}{\sqrt{N}} = C\sqrt{N} = C\sqrt{\phi A}, \tag{10.4}$$

where ϕ is the photon fluence (e.g., photons per unit area) used to form the image. Equation 10.4 is the basic mathematical form of the Rose model, sometimes called the "Rose model equation."

Rose found that human observers require an SNR (k) of 5–7 to visualize a low-contrast target as separate from its background. *The value of "k" is used to establish the threshold of visual detection or perception.* This value was derived experimentally based on the work of Blackwell that we discussed earlier. In this course, we will use $k = 5$ in examples and problems.

An important use of the Rose model equation is to estimate the radiation exposure necessary to visually perceive a small low-contrast object.

Example 10.1: Exposure Calculation Based on Rose Model

Assume that there is an air bubble in a 20 cm thick tank of water (Figure 10.2). Calculate the incident photon fluence and exposure needed to just see a 1 cm diameter air bubble as a function of photon energy E.

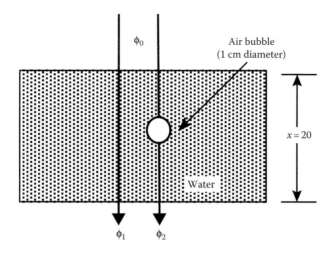

FIGURE 10.2 A 1 cm diameter air bubble contained in a 20 cm thick volume of water.

Solution

The Rose model in Equation 10.4 leads to the following equation for k^2:

$$k^2 = c^2 \Phi_1 A, \tag{10.5}$$

where
   k is the constant with value of 5
   C is the contrast for air bubble
   Φ_1 is the number of photons per unit area to form the image (in background)
   A is the area of air bubble
   d is the bubble diameter

If Φ_0 is the photon fluence without the water phantom, then the photon fluence through the water (background) is

$$\Phi_1 = \Phi_0 e^{-\mu x}. \tag{10.6}$$

Assuming no attenuation by the air bubble, the photon fluence through the water and the air bubble of diameter d (target) is

$$\Phi_2 = \Phi_0 e^{-\mu(x-d)}. \tag{10.7}$$

The air bubble is therefore imaged with a contrast of

$$C = \frac{\Phi_2 - \Phi_1}{\Phi_1} = \frac{e^{-\mu(x-d)} - e^{-\mu x}}{e^{-\mu x}} = e^{\mu d} - 1. \tag{10.8}$$

An air bubble of diameter d has an area A of

$$A = \frac{\pi d^2}{4}. \tag{10.9}$$

From the Rose model, the area also can be calculated combining Equations 10.5, 10.6, and 10.8, and assuming that $k = 5$,

$$A = \frac{k^2}{C^2 \Phi_1} = \frac{25}{\left(e^{\mu d} - 1\right)^2 \Phi_0 e^{-\mu x}}. \tag{10.10}$$

Equating 10.9 and 10.10, we see that the incident photon fluence to just see the air bubble is

$$\Phi_0 = \frac{100}{\pi d^2 \left(e^{\mu d} - 1\right)^2 e^{-\mu x}} \left[cm^{-2} \right], \tag{10.11}$$

where
    d is the diameter of air bubble (cm)
    x is the thickness of water tank (cm)
    μ is the linear attenuation coefficient of water (cm^{-1})

We can determine an equivalent radiation exposure X corresponding to photon fluence Φ_0 using the following equation from Johns and Cunningham, converted into the appropriate units for this problem:

$$X = 1.833 \times 10^{-8} \Phi_0 \cdot E \cdot \left(\frac{\mu_{en}}{\rho}\right)_{air} (mR). \tag{10.12}$$

Here, E is in units of keV, Φ_0 in units of cm^{-2} and the mass-energy absorption coefficient of air is in units of cm^2/g. So the exposure required for the air bubble of diameter d in a water phantom of thickness x is

$$X = \frac{1.833 \times 10^{-6} E \left(\dfrac{\mu_{en}}{\rho}\right)_{air}}{\pi d^2 (e^{\mu d} - 1)^2 e^{-\mu x}} (mR). \tag{10.13}$$

The minimum input exposure required to see 1 cm and 1 mm diameter air bubbles in a 20 cm thick water tank is given for different photon energies in Table 10.1. Note that the minimum exposure occurs at 80 keV for both bubble diameters, and that a much larger exposure ($\times 10^4$) is needed to see the smaller bubble (1/10th diameter).

TABLE 10.1   Input Radiation Exposure to Just See Air Bubble as a Function of Photon Energy (20 cm Water Phantom). Bold indicates lowest input exposure

Photon Energy (keV)	$\left(\dfrac{\mu_{en}}{\rho}\right)_{air}$ (cm²/g)	$\left(\dfrac{\mu}{\rho}\right)_{water}$ (cm²/g)	Input Exposure (mR) (1 cm bubble)	Input Exposure (mR) (1 mm bubble)
10	4.533	5.066	1.10E+35	6.1E+41
15	1.242	1.568	3.10E+07	1.6E+12
20	0.4942	0.7613	18.15	3.78E+05
30	0.1395	0.3612	0.0177	2.48E+02
40	0.0625	0.2629	0.0031	39.49
50	0.0382	0.2245	0.0016	19.27
60	0.0289	0.2046	0.0012	14.17
80	**0.0236**	**0.1833**	**0.0011**	**12.58**
100	0.0231	0.1706	0.0012	13.81
150	0.0249	0.1505	0.0017	19.23

10.3 CONTRAST-DETAIL ANALYSIS

A number of investigators, most notably Gerald Cohen, developed an experimental technique based on the Rose model to evaluate object detectability at the threshold of human visibility in medical images. This method, called contrast-detail curve analysis, is based on numerous assessments at different contrasts and object sizes, and the contrast required for visualization is plotted as a function of object diameter. The plotted curve is called a contrast-detail curve. Theoretical sketches of several contrast-detail curves are provided in Figure 10.3 and support our intuitive notion of the relationship between contrast and object size in a medical image. Large objects can be visualized at low contrast (lower right), while small objects require high contrast (upper left). The contrast-detail curve declines asymptotically toward the lower right corner (low-contrast, large objects). A contrast-detail curve allows one to predict the threshold contrast needed for detection of a wide range of object sizes.

Systems can be compared using their contrast-detail curves (Figure 10.4). One curve can be generated from each system, or on the same system under different operating conditions. For example, a contrast-detail curve can be generated at one radiographic technique (i.e., kVp, mA, and exposure time), and compared with another curve using a different technique. *The lower curve indicates a better technique*, one that supports detection with lower object contrast (System A in Figure 10.4).

10.3.1 Experimental Determination of Contrast-Detail Curve

A contrast-detail curve can be generated by a panel of observers who attempt to detect simple objects in a test image from the system under investigation. We will describe how a test object can be designed for film-screen radiography. Of course, the design of an appropriate test object will depend on the system being evaluated, but the general principles follow those presented.

A typical test x-ray object for contrast-detail analysis, called a Rose model phantom, is shown in Figure 10.5. The phantom is a wedge-shaped piece of plastic (or some other material having low attenuation) through which holes of various sizes are drilled. The plastic wedge is

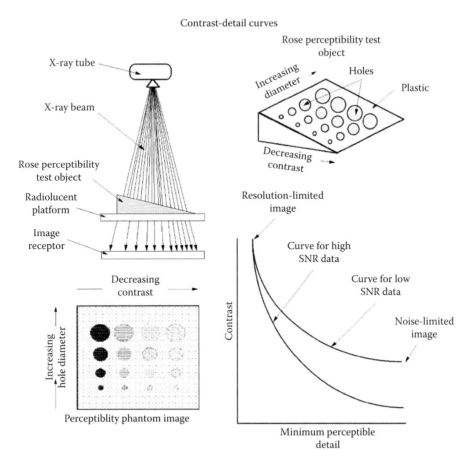

FIGURE 10.3   Contrast-detail curves are obtained by having observers detect circular targets in a radiograph of a Rose model phantom. Resulting contrast-detail curves indicate the contrast needed to minimally perceive objects of increasing size for high and low x-ray fluence (high and low SNR).

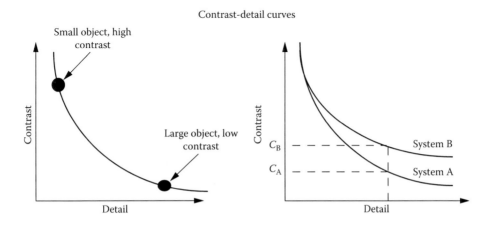

FIGURE 10.4   Two imaging systems can be compared using contrast-detail curves. For a given object size ("detail"), system A can detect the same level of detail at a lower contrast than system B, indicating that system A may be superior to system B.

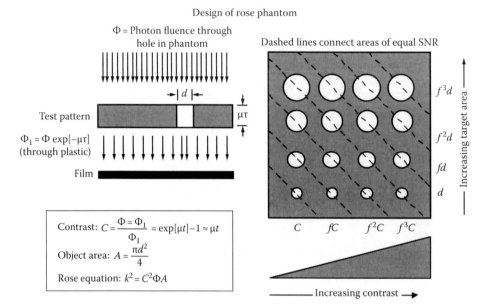

FIGURE 10.5 In designing a Rose model phantom, objects of equal SNR are determined by increasing the hole diameter and decreasing the material thickness by the same factor (f). These fall along the diagonal lines.

placed on a film-screen cassette and imaged, and an observer is asked to report which holes are just perceptible in each column of the resulting image. The plastic thickness provides varying contrast and hole diameters provide varying detail. A contrast-detail curve is generated by graphing contrast versus hole diameter for minimally perceptible holes.

For construction, assume we have a piece of plastic of thickness t through which a hole of diameter d is drilled. The plastic is radiographed with the x-ray beam parallel to the axis of the drilled hole. If Φ is the photon fluence (photons/cm²) through the hole and Φ_1 the fluence through an adjacent section of plastic, the contrast between the hole and the plastic is

$$C = \frac{\Phi - \Phi_1}{\Phi_1} = e^{\mu t} - 1, \tag{10.14}$$

where μ is the linear attenuation coefficient of the plastic. In "low-contrast" situations (i.e., where $\mu t \ll 1$), the exponential term can be expanded as a Taylor's series $\exp(\mu t) \approx 1 + \mu t$ so that contrast is approximately proportional to the plastic thickness

$$C \approx \mu t. \tag{10.15}$$

The hole is projected onto the radiograph as a circle having diameter d (without magnification). Therefore, the area of the object we wish to detect is

$$A = \frac{\pi d^2}{4} \tag{10.16}$$

and for low-contrast conditions, the photon fluence Φ through the hole is approximately equal to that through the plastic (background). Therefore, the Rose model states that

$$k^2 = C^2\Phi A = (\mu t)^2 \, \Phi \cdot \frac{\pi d^2}{4}, \tag{10.17}$$

where k is the SNR required by the observer to detect the holes in the radiograph. (For observers under ideal conditions with threshold detection levels, k has a value from 5 to 7.)

The Rose model phantom is typically constructed with plastic thickness (i.e., contrast) increasing in one direction and hole diameter (i.e., detail) changing in the perpendicular direction. The threshold contrast and thickness are related by

$$C^2 = (\mu t)^2 \tag{10.18}$$

and the hole area is related to its diameter by

$$A = \frac{\pi d^2}{4} \tag{10.19}$$

If the hole diameter d is increased by a multiplicative factor while the phantom thickness t is decreased by the same factor, then k^2 remains constant (Equation 10.17). This means that on the Rose phantom, a line connecting holes at increasing diameter with corresponding decreasing contrast has a fixed value of k^2 and represents a diagonal line along which the SNR k is constant for a given photon fluence (Figure 10.5). *Hence, a radiograph of the Rose phantom will contain a hypothetical diagonal line along which threshold detection occurs for all targets.* Targets at larger size and contrast on one side of this line should be detected more easily. Those on the other side of the line with lower contrast and size should be more difficult to detect.

Example 10.2: Rose Model Phantom Design

Design a Rose phantom to be used in conventional film-screen radiography. The phantom will be radiographed with an exposure of 0.1 μR at an effective energy of 20 keV. Assume that the phantom is made of water-equivalent plastic.

Solution

Obviously, we have some latitude in the design of the phantom. Let us start by calculating the plastic thickness required to see a 2 mm diameter hole. We know that at energy of 20 keV, and given that

$$\left(\frac{\mu_{en}}{\rho}\right)_{air} = 0.4941 \, \text{cm}^2/\text{g} \tag{10.20}$$

the photon fluence for an exposure $X = 0.1$ μR is

$$\Phi = \frac{5.434 \times 10^7 \left[\dfrac{\text{keV}}{\text{g-mR}} \right] \cdot X[\text{mR}]}{\left(\dfrac{\mu_{en}}{\rho} \right) \left[\dfrac{\text{cm}^2}{\text{g}} \right] \cdot E \left[\dfrac{\text{keV}}{\text{photon}} \right]} = 551 \text{ photons/cm}^2 \tag{10.21}$$

From the Rose model and Equation 10.17,

$$k^2 = C^2 \Phi A = (\mu t)^2 \, \Phi \left(\frac{\pi d^2}{4} \right) \tag{10.22}$$

and using a value of $k = 5$,

$$t = \frac{2k}{\mu d \sqrt{\pi \Phi}} = 1.58 \text{ cm}, \tag{10.23}$$

where $\mu = 0.7613$ cm^{-1} is the linear attenuation coefficient of the water-equivalent phantom material at 20 keV.

We have now established a single thickness and hole diameter for the phantom. From our preceding discussion, we know we can increase the phantom thickness by a multiplicative factor while we decrease hole diameter in the opposite direction by the same factor. *Experience has shown that this factor should be no larger than* $\sqrt{2}$. Using this factor, we will select hole diameters of

   1 mm, 1.41 mm, 2 mm, 2.83 mm, 4 mm

and phantom thickness of

   0.79 cm, 1.12 cm, 1.58 cm, 2.23 cm, 3.16 cm.

Both change stepwise by a factor of $\sqrt{2}$. This leads to a phantom with five rows each at one of the designed thicknesses. Within each row, there will be five holes, one for each of the five diameters.

10A APPENDIX

A contrast-detail curve is a graph of contrast versus diameter where these are the values at the threshold of visual perception. The data are generally taken from several volunteers who view images of a Rose model type phantom. The basic Rose model equation can be used to model a contrast-detail curve after modifications to correct for differences due in asymptotic behavior for large and small diameter objects (Figure 10.4).

The solution of the basic Rose model equation in terms of contrast versus diameter is as follows:

$$C = \frac{2k}{\sqrt{\phi\pi}} \cdot \frac{1}{d},$$ (10A.1)

where

    k is the SNR for minimal perception
    d is the diameter of the object (cm)
    ϕ is the photon fluence (cm^{-2})

The asymptotic behavior for the basic model tends toward infinite contrast with small d and zero contrast for large d. Neither cases are possible, nor is the behavior correct. We see an indication of the actual asymptotic behavior in Figure 10.4 for both small and large.

Contrast asymptote correction: The most straightforward correction to the basic Rose model is to add a baseline contrast (C_0) that must exist even for larger diameter objects. This leads to the following modified Rose model:

$$C = \frac{2k}{\sqrt{\phi\pi}} \cdot \frac{1}{d} + C_0.$$ (10A.2)

The departure from the basic Rose model asymptote has been suggested to be a problem with the human visual system for objects exceeding the diameter of the eye's high visual acuity region, the foveal region. The added constant partly compensates for this and C_0 has been determined experimentally from studies of contrast-detail curves.

Diameter asymptote correction: The departure from the basic Rose model at small d has been modeled as a geometrical limitation of the imaging system, where images do not correctly reproduce small diameter objects. The correction is to add a small diameter δd leading to the corrected Rose model:

$$C = \frac{2k}{\sqrt{\phi\pi}} \cdot \left(\frac{1}{d + \delta d} \right) + C_0.$$ (10A.3)

Equation 10A.3 is the asymptotically corrected version of the Rose model equation, a more correct and detailed version.

The theory behind the correction in diameter follows from the need of the basic Rose model equation to accommodate for the spatial resolution of the system. How the increase in area is modeled as an increase in diameter is given below:

Partly correct Rose model:

$$C = \frac{k}{\sqrt{\phi A}} + C_0.$$

Here, A is increased by δA to model system blurring:

$$C = \frac{k}{\sqrt{\phi(A+\delta A)}} + C_0 = \frac{k}{\sqrt{\phi A'}} + C_0.$$

For a circular cross-sectional object:

$$A' = \frac{\pi(d+\delta d)^2}{4}.$$

Substitution into the equation for C yields

$$C = \frac{k}{\sqrt{\phi\dfrac{\pi(d+\delta d)^2}{4}}} + C_0.$$

Simplifying leads to

$$C = \frac{2k}{\sqrt{\phi\pi}} \cdot \frac{1}{d+\delta d} + C_0.$$

It is instructive to review the effect of each parameter in Equation 10A.3 on minimal contrast:

k—(larger k means more contrast needed for all diameters) affected by environment: brightness of viewer, background light.

C_0—(larger C_0 means more contrast needed but effect greatest for larger diameters) complexly interrelated with image size projected onto the retina.

δd—(larger δd effectively increases size of object shifting the contrast-detail curve toward larger diameters) related to spatial resolution (effective diameter of the PSF) includes eye resolution for visual tasks.

ϕ—(larger ϕ means less contrast needed for all diameters). For x-ray systems, larger photon fluence is possible with slower film-screen systems. For digital imaging, plates can vary over several orders of magnitude to see smaller low-contrast objects if needed.

HOMEWORK PROBLEMS

P10.1 A patient is suspected of having a tumor located in his lung. Assume that in the area of the lung, the body thickness is equivalent to 10 cm of water and that the soft tissue has an attenuation coefficient corresponding to an HVL of 3 cm H_2O.

(a) Calculate the entrance exposure required to detect a 1 cm tumor (water-equivalent) sitting in the lung. For simplicity, assume relative path lengths of 10 and 11 cm rather than worrying about the replacement of 1 cm of low-density lung tissue.

TABLE 10.2   Attenuation Data for Homework Problem P10.4

Photon Energy (keV)	$\left(\dfrac{\mu_{en}}{\rho}\right)_{air}$ (cm²/g)	$\left(\dfrac{\mu}{\rho}\right)_{water}$ (cm²/g)	$\left(\dfrac{\mu}{\rho}\right)_{Iodine}$ (cm²/g)
10	4.533	5.066	162.7460
15	1.242	1.568	54.5977
20	0.4942	0.7613	25.0943
30	0.1395	0.3612	8.4212
40	0.0625	0.2629	22.3827
50	0.0382	0.2245	12.5016
60	0.0289	0.2046	7.6938
80	0.0236	0.1833	3.5496
100	0.0231	0.1706	1.9562
150	0.0249	0.1505	0.6984

(b) Repeat the calculation for a 1 mm tumor.

(c) Repeat (b) assuming path lengths of 15.0 and 15.1 cm (i.e., a 15 cm thick body region) if scatter is neglected.

P10.2 Compare the x-ray exposure required for noise-limited radiographic imaging for a small cubic object of dimension a and for a cube of similar material but dimension $2a$.

P10.3 For a fixed radiographic entrance exposure, how is the SNR of an image affected if the diameter of the object is halved and its contrast tripled.

P10.4 We wish to determine the incident exposure required to image a 1 mm diameter, 1 mm long arterial stenosis that is located in 20 cm of water-equivalent tissue. Assume that the artery contains 10 mg/cm² of iodine at a density of 2 g/cm³. Calculate the incident exposure for the energies listed in Table 10.2. Using this information, determine the optimal energy for imaging the stenosis.

P10.5 The Rose model phantom given in Example 10.2 is used in various imaging situations.

(a) Draw a contrast-detail curve for this phantom at the imaging technique (0.1 µR per exposure, 20 keV effective energy) given in Example 10.2.

(b) Draw a contrast-detail curve if the exposure time is doubled while the exposure rate is maintained at the same level given in (a).

(c) If the effective energy of the beam is changed from 20 to 30 keV, but the exposure is still 0.1 µR, how does this change the contrast-detail curve? Answer this question quantitatively.

(d) If the thickness of the phantom is doubled, how does this change the contrast-detail curve obtained in (a)?

(e) The Rose phantom is used with a fluorographic system where the electronic noise equals the quantum statistical noise. How does this change the contrast-detail curve obtained in (a)?

(f) The Rose phantom is used in a situation where the scatter fraction is 75%? How does this change the contrast-detail curve from that given in (a)?

CHAPTER 11

Receiver Operating Characteristic (ROC) Analysis

11.1 INTRODUCTION

Unlike imaging system performance testing methods that focus on spatial resolution, contrast, and noise, the approach used in receiver operating characteristic (ROC) analysis focuses on outcomes. ROC analyses compare the result of a diagnostic imaging test (with decision positive or negative) with the known clinical condition (disease present or absent). Both the test decision and disease condition are restricted to binary results (+/−), so there are four possible outcomes (Table 11.1).

Two of the test outcomes are correct or true (TP & TN) and two are incorrect or false (FP & FN). A perfect test would yield only correct or true outcomes. If the diagnostic or decision threshold is fixed, we can evaluate groups of patients with disease (Disease+) and without disease (Disease−) to determine values for Table 11.1. The distributions of test results for patients with and without a disease reveal that these can overlap appreciably (Figure 11.1). In this figure, the threshold for a positive test was set at the point where the two distributions crossed to emphasize this overlapping. While in practice, we would not have sufficient patient data or time to determine such distributions, this figure helps to illustrate several important concepts. In practice, we evaluate a limited number of patients with and without the disease and based on a threshold for positive test determine the number of patients in each of the groupings, indicated by the regions labeled as TP, FP, TN, FN in Figure 11.1. From these data, we can fill in Table 11.1.

11.2 TEST CHARACTERISTICS

There are several indices used to gauge performance of diagnostic tests.

Sensitivity: The sensitivity of a test, also called the true-positive fraction (TPF), is an index of how well the test performs with patients who have the disease ($0 \leq \text{TPF} \leq 1$). Test sensitivity = TP/(TP + FN), where TP is the number of patients with the disease

TABLE 11.1   Test Result vs. Disease Condition

Test	Disease	
	Present (*D*+)	Absent (*D*−)
Positive (T+)	True positive (TP)	False positive (FP)
Negative (T−)	False negative (FN)	True negative (TN)

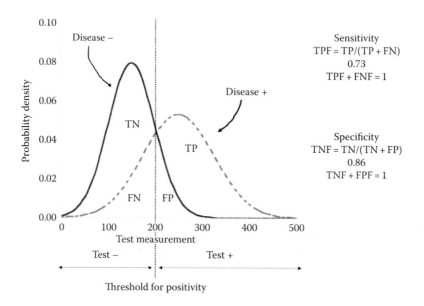

FIGURE 11.1 Probability distributions of test results for disease and nondiseased patients. Calculated sensitivity and specificity are for the test threshold indicated by the vertical line.

that were above the test threshold and FN the number with the disease that were below the threshold. *Test sensitivity provides a measure of the accuracy of the test with a disease population.*

Specificity: The specificity of a test, also called the true-negative fraction (TNF), is an index of how well the test performs with patients from the nondiseased population ($0 \leq \text{TNF} \leq 1$). Test specificity is calculated using the number of true-negative and the false-positive patients (TN & FP) where specificity = TN/(TN + FP). *Test specificity provides a measure of the accuracy of the test with a nondiseased population.*

Accuracy: The overall accuracy of a test is the fraction of correct test results or diagnoses for both disease and nondiseased populations. Test accuracy is calculated as the number of patients with correct test results divided by the number in the group, so test accuracy = (TP + TN)/(TP + FP + TN + FN).

Prevalence: The prevalence of the disease is calculated as the fraction of patients who have the disease, so disease prevalence = (TP + FN)/(TP + FP + TN + FN).

Inspection of Figure 11.1 reveals that the test indices vary as position or width of the two distributions changes or if a different test threshold is used. There is therefore a need to characterize a test (imaging exam) with a more comprehensive outcome measure. ROC analysis attempts to do this and has gained popularity as a general method for determining the value of a diagnostic imaging system to test for disease. I will present a brief introduction to the theory and methods used for ROC analysis. For those that want additional details, I recommend the excellent text by Swets and Pickett, *Evaluation of Diagnostic Systems— Methods from Signal Detection Theory.*

11.3 ROC ANALYSIS

ROC analysis is used to assess the accuracy of diagnostic outcomes for an imaging system as a function of varying decision or test threshold. The analysis is done without the explicit use of resolution, contrast, or noise measures, though outcomes clearly involve these. Unlike contrast detail analysis with a Rose model phantom, ROC analysis can be used to compare different imaging modalities (MRI, x-ray CT, SPECT, or PET) for outcomes with the same disease. Additionally, ROC analysis can be used for a single imaging modality as an aid to optimize acquisition parameters, such as different TR and TE for MRI and kVp for CT for a specific disease. *Finally, even different forms of image analysis can be evaluated for specific diseases.* The main obstacle with the ROC approach is that we must know whether the disease is present or absent, the objective of diagnostic imaging, and we may not know this until confirmed by multiple diagnostic exams or perhaps biopsy. This is clearly a problem for prospective studies where we seek to detect the presence of disease.

In ROC analysis, the *true-positive fraction* [TPF = TP/(TP + FN)] of a diagnostic exam is plotted against *false-positive fraction* [FPF = FP/(FP + TN)] as the diagnostic threshold is varied (Figure 11.2). If a physician were trying to determine the existence of a tumor on a chest film, TPF would be the fraction of times the physician said there was a tumor (test was positive) when there actually was a tumor in the patient. Similarly, the false-positive fraction is the fraction of times the physician said there was a tumor when no tumor was present. As trivial examples, if a physician always said that a tumor was present in a chest film, then he would call all the examples of tumors correctly in which case his TPF would be 100%. Unfortunately, he would diagnose all healthy people as having a tumor, and his false-positive fraction would be 100% (upper right extreme of ROC curve). Alternatively, if a physician always said that no tumor existed, his TPF would be 0%, but he also would miss all the tumors on the film, so his false-positive fraction would be 0% (lower left extreme of ROC curve).

An ideal imaging system would give no false positives unless the observer insisted upon calling everything positive (No overlap of *D*+ and *D*− distributions). Its ROC curve therefore would hug the upper left corner and top of the graph range. On the other hand, if the image conveyed no information and the observer was forced to guess whether or not the object was present, the ROC curve would be a diagonal line from the lower left to the upper right corner (useless test is a chance line, Figure 11.3). As shown in this figure, the amount by which the ROC curve bows away from the diagonal and toward the upper left-hand corner is an indication of the usefulness of the imaging technique.

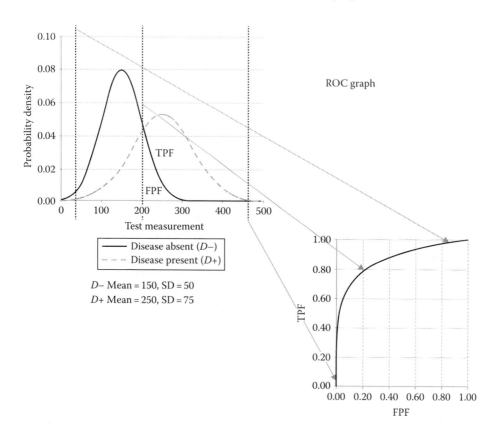

FIGURE 11.2 Graphing an ROC curve. Three diagnostic thresholds are highlighted as dotted lines.

11.4 COMPARING ROC CURVES

There are several measures that can be used for comparing ROC curves. One is the minimum distance from the curve to the point indicated by a perfect system (TPF = 1; FPF = 0 at the upper left corner of the graph). A more comprehensive measure is the area under the ROC curve. For a perfect test, this area = 1.0; for the useless test, the area = 0.5. It is therefore the area in excess of 0.5 that is important. The area measure is considered superior to the minimum distance measure, since it provides an assessment over a range of diagnostic thresholds for the D+/– distributions. There are cases where the curves cross when comparing modalities or methods and the one with the smallest distance may not be preferred across the full range. At other times, the FPF determines the operating point and then the ROC curve with the largest corresponding TPF would indicate the better system.

11.5 SELECTING OPERATING POINT

As indicated earlier, the test criteria for positivity could be set such that the physician is always correct for the D+ patients, allowing treatments to be planned. It is definitely important to achieve a high TPF; however, as can be seen from Figures 11.1 through 11.3, for any practical imaging exam adjusting test criteria to increase the TPF will also increase the FPF. Likewise, adjusting the threshold to reduce FPF also reduces TPF. As with all

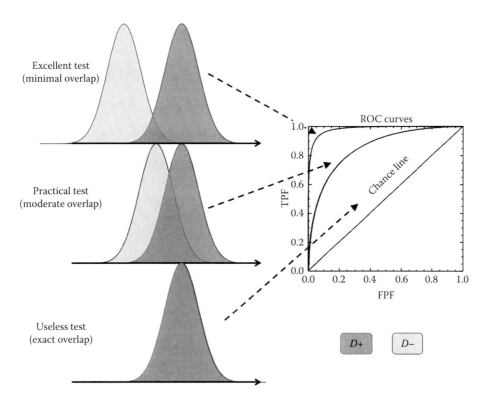

FIGURE 11.3   ROC curves for three tests, one considered excellent with minimal overlap of D+ and D− distributions, one practical, and one useless where the two distributions are identical.

medical decision-making, careful consideration must be given to the balance between having a high TPF and a low FPF.

If we lower the test threshold, we increase the number of FP patients and they are ultimately faced with additional diagnostic exams and increased medical expense. If we increase the threshold, we can miss patients with the disease that we might have discovered during an early stage where it might be more successfully and economically managed. This is the dilemma we are faced with in medical decision making. Each facility must determine the decision threshold that best fits with their patient population and imaging capabilities.

11.6 AN ROC EXAMPLE

ROC analyses use knowledge of the absence or presence of disease, which might not be available when we need to evaluate an imaging system. However, as is often done with resolution, contrast, and noise testing, we can substitute a phantom for the patient with and without lesions (or objects simulating lesions). While this is not optimal, the phantom approach provides a means to compare systems until patient outcomes are available.

For visual diagnoses, it is more difficult to assess and vary diagnostic threshold than those that have quantitative measures. A common ROC method for visual diagnoses is to have observers inspect an image and rate their confidence that an object is present and then check this decision against a "gold standard" (known presence or absence of the object) to determine true-positive and false-positive fractions. It is common to use a rating system that ranges from "0" when an observer is least confident that an object is present through "4" when an observer is most confident that the object is present.

Assume that observers used this rating system to indicate whether or not objects are present in a radiograph (Table 11.2).

Here, observer rating "4" represents the highest positive threshold (certain of presence) and rating "0" the lowest positive threshold (nearly certain not present). Points on the ROC curve are obtained from cumulative results, because as the threshold is lowered from "4" toward "0," the lower threshold would also include those with a higher threshold, and ultimately, the lowest threshold would include all. Table 11.3 is the worksheet used to determine TPF and FPF for the example.

Note that the ROC analysis does not require the same number of observations for the $D+$ and $D-$ groups, since the objective was to estimate only fractions; however, the same observers were used. This table provided pairing for TPF and FPF for each threshold, and these paired data can be plotted as an ROC curve (Figure 11.4).

TABLE 11.2   Data for ROC Analysis Example

For images where the *object is present*, the observers give the following scores:

Observer Rating	Number of Decisions
0—Nearly certain not present	2
1—Probably not present	3
2—Uncertain	7
3—Probably present	5
4—Nearly certain present	3

For images where the *object is not present*, the observers give the following scores:

Observer Rating	Number of Decisions
0—Nearly certain not present	5
1—Probably not present	7
2—Uncertain	8
3—Probably present	3
4—Nearly certain present	2

TABLE 11.3   Calculation of TPF and FPF for an Example for ROC Analysis

D− Objects Absent				D+ Objects Present			
t	$N(t)$	Cumulative	FPF	t	$N(t)$	Cumulative	TPF
4	2	2	0.08	4	3	3	0.15
3	3	5	0.20	3	5	8	0.40
2	8	13	0.52	2	7	15	0.75
1	7	20	0.80	1	3	18	0.90
0	5	25	1.00	0	2	20	1.00

t is the observer rating level from Table 11.2.

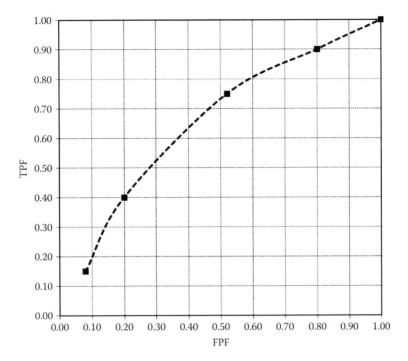

FIGURE 11.4   ROC curve for the example data from Tables 11.2 and 11.3.

Though only five points were used in this ROC study, we could include another point at (TPF = 0; FPF = 0) if needed, as this would be the threshold that had no positive test results. The dashed lines are provided to better illustrate the curve trend for the thresholds used. This ROC result is not an example of what might be considered a good imaging system, since the curve is far from the upper left corner of the graph (ideal test); rather, it is provided as an example of how to acquire data, process results, and graph an ROC curve. The estimated area for the curve is 0.644, which is quite low, since random guessing would have an area of 0.500.

Incidentally, if you assumed the opposite ordering of ratings for the test (0-to-4 vs. 4-to-0) you would get a much different ROC curve, one below the diagonal and with different values for paired TPF and FPF. This is a good way to check that you ordered your data correctly.

HOMEWORK PROBLEM

P11.1 Use Excel to create an ROC curve where D+ (with disease) and D− (without disease) distributions of test results are modeled as Gaussian functions. Use a range of test results from 1 to 100. The mean and FWHM for the Gaussian distributions are as follows:

	D−	D+
Mean	45	55
FWHM	20	25

What is the area for this ROC curve?

IV

Dynamic Imaging

CHAPTER 12

Digital Subtraction Angiography

12.1 INTRODUCTION

In this chapter, we describe a technique known as digital subtraction angiography (DSA), which is also referred to as digital radiography, digital fluoroscopy, and photoelectronic imaging. We will use these terms somewhat interchangeably though the term "digital subtraction angiography" refers specifically to techniques that subtract two images, one obtained before and one after the contrast media is administered to the patient, for the purposes of studying blood vessels (angiography). The more general term, digital radiography, encompasses the use of all digital electronic techniques in x-ray imaging. According to some writers, this term also includes the use of x-ray computed tomography (CT), though digital radiography in this chapter will refer only to those techniques in which digital systems are used to acquire planar rather than tomographic images. We will concentrate on a hybrid DSA system that uses an image intensifier viewed by video camera prior to digitization, since these analog components provide excellent examples.

12.2 DESIGN FOR A HYBRID DSA SYSTEM

The generic diagram of a hybrid digital radiographic system in Figure 12.1 shows many interacting components. The heart of this system is a digital processing component that acquires images from a video camera based on timing signals for the x-ray generator and analog-to-digital convertors (ADCs). The operator component provides human control of the flow of data from the x-ray source throughout the system.

Image acquisition begins when timing signals, delivered to the x-ray generator under computer control, initiate the production of x-rays that are transmitted through the patient and received by the image intensifier. An aperture, placed between the image intensifier and the video camera, controls the amount of light delivered to the camera. The aperture is adjusted to optimize the signal-to-noise ratio (SNR) of the acquired image, as will be discussed later in this chapter. A video camera receives the light image from the image

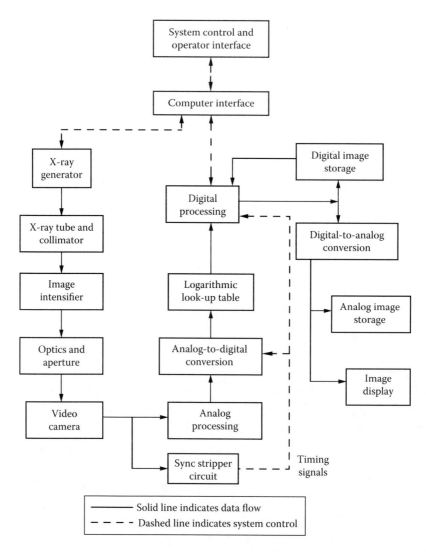

FIGURE 12.1 Box diagram of components in a hybrid DSA system.

intensifier and converts it to an electronic signal that is delivered to the image processor as analog video. The image processor digitizes the image, stores it in memory, and makes it available in digital format for subtraction from other images acquired at a different time or at a different energy. The basic components of the DSA system including the x-ray tube and x-ray generator, the image intensifier, and the video camera are similar to but must be of *higher quality than those used in an analog fluoroscopy system* to ensure proper synchronization and match between analog and digital components.

A common image-processing algorithm used with digital radiographic systems is temporal subtraction (Figure 12.2). For this technique, dynamic images of the patient are acquired at a rate of one exposure per second or higher. A contrast agent is injected into the patient either intravenously or intra-arterially. A set of dynamic images is acquired as the contrast agent flows into the area of interest. The difference between contrast images

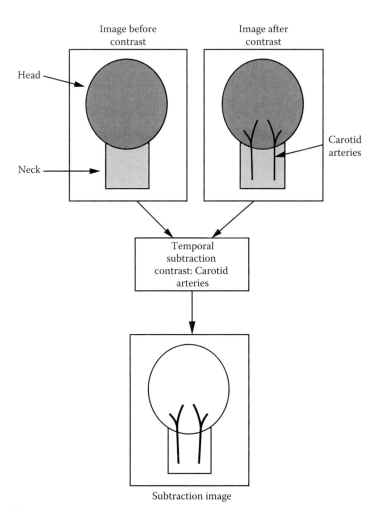

FIGURE 12.2 Temporal subtraction to highlight change due to the contrast material in carotid artery.

and "mask" images (without contrast) is used to isolate the contrast signal. This removes static (nonmoving) anatomical structures common to both. The elimination of static structures makes small arteries visible in the subtraction images that were not visible or barely visible before subtraction.

12.3 CLASSICAL ANALOG IMAGING SYSTEM COMPONENTS

12.3.1 Image Intensifier

The subtraction algorithm assumes that the patient's anatomy is similar or identical in both the contrast and mask images. The video camera, the x-ray tube, and the other system components must be stable to ensure this for anatomical structures to be properly removed by subtraction. To preserve the contrast available in the radiographic image, the image intensifier must have a *high contrast ratio and dynamic range*. Analog-to-digital conversion should provide sufficient spatial sampling to preserve the resolution of the image

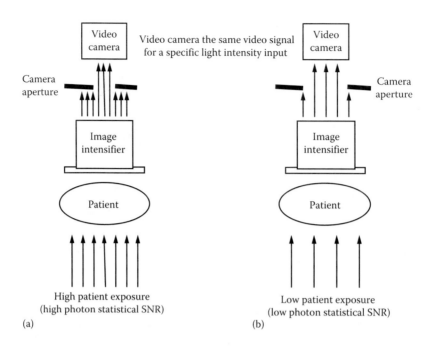

FIGURE 12.3 The effect on patient exposure by the aperture used to control light level for an image intensifier system: (a) small camera and (b) large camera apertures.

intensifier, sufficient temporal sampling to capture the time-varying video signal, and sufficient dynamic range.

12.3.2 Light Aperture

A light aperture (Figure 12.3), similar to those found on single-lens reflex cameras, is placed behind the output phosphor of the image intensifier to control the light level reaching the video camera for a given exposure rate. A small aperture requires a greater radiation exposure to deliver the proper light level to the video camera, decreasing the effect of quantum noise and producing a better overall SNR in the image. Conversely, a large camera aperture is used when the objective is to reduce patient exposure in cases where quantum noise does not limit the diagnostic information in the image.

12.3.3 Video Camera

A key component in the imaging chain of a hybrid DSA system is the analog video camera. The basic function of the video camera is to produce an analog electronic signal that is proportional to the light intensity incident at the video target of the camera (Figure 12.4).

The photoactive element is the video target that changes in electrical conductivity when exposed to light. Scanning an electron beam across sequential lines of the video target creates the video signal. Regions where the target is exposed to high light intensities have high conductivity, and thus a large current. Regions of the target exposed to low light intensities have low conductivity and thus a small current. The resulting signal is a measure of the light intensity exposing the video target. The information is read out serially as

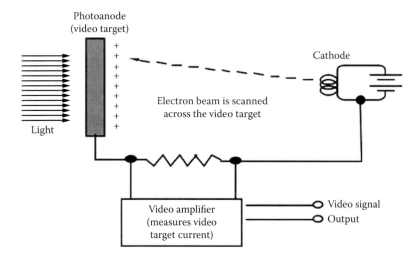

Photoanode
(video target)

Cathode

Electron beam is scanned
across the video target

Light

Video amplifier
(measures video
target current)

Video signal

Output

FIGURE 12.4   Schematic of video camera.

the electron beam is swept over the target to generate the analog video signal. This reading process resets the conductivity of the video target. The video signal is a time-varying signal that encodes the two-dimensional light image at the target as an electrical temporal record. Time points in the video signal correspond to spatial locations of the light image at the target.

The video camera's target can be read out in one of two ways. In 525-line video cameras used in the broadcast industry, the electron beam is scanned across the target in 262.5 passes across the area of the target. The resulting 262.5 lines encode the image of the target in what is called a video field (Figure 12.5). A video field is produced every 1/60 of a second. During alternating fields, the electron beam scans lines between the lines of the other field. Therefore, two fields are acquired using an interlaced scanning pattern. The fields are called "odd" and "even" with each field comprising 1/60 of a second, and two fields make one "frame" acquired in 1/30 of a second that is comprised of 525 lines. This frame is the full image, and new images are acquired at a rate of 30/s. The interlace scan mode was chosen by the early broadcast industry to reduce bandwidth during transmission while avoiding flicker in the viewed video image.

Interlaced scanning is not ideal for digital radiography. The basic problem with the interlaced mode is that the video fields are not read out in one pass. An alternate continuous scanning mode scans lines sequentially such that the entire frame is filled in one pass. To distinguish this mode from interlaced video, it is called noninterlaced video.

Most video targets have a small amount of lag so that even when they are exposed to a uniform abrupt change in light level, several fields are needed to adjust (Figure 12.6). Also, assuming that beam scanning and x-ray exposure start at the same time, those locations scanned later in the frame will be exposed longer than those scanned early, and it takes several video fields before the output signal is stable. Thus, just after the x-ray beam is initiated, these effects lead to images with incorrect signals. The early fields must be discarded, although this is clearly undesirable, since it underutilizes the x-ray exposure delivered to the patient.

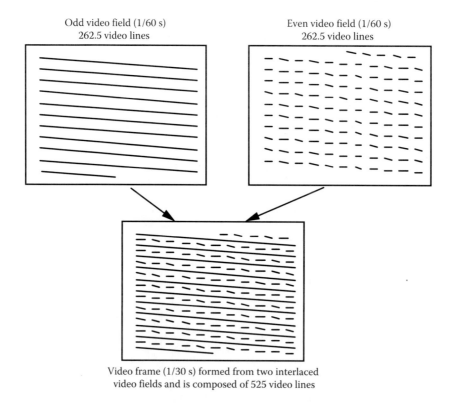

FIGURE 12.5 Layout of even and odd fields for interlaced video scanning.

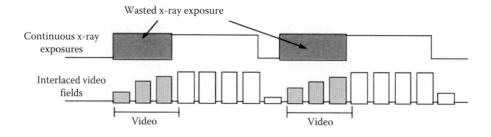

FIGURE 12.6 Problem with x-ray exposure for the interlaced scan mode.

An alternative to the continuous scanning mode called progressive scanning resolves the problems with wasted exposure (Figure 12.7). When this mode is used, an image is stored in the target during a short x-ray exposure without scanning and is then fully scanned after the x-ray beam is turned off. This approach eliminates the wasted x-ray exposure seen for continuous x-ray exposure scanning. It does this by separating exposure and readout times.

There are other aspects of the video camera important in a digital radiographic system. First, the magnitude of the video signal should be directly proportional to the input x-ray fluence, that is, linear response. Second, the video camera must have low lag. This means that an image acquired at one point in time by the video camera should not persist on

FIGURE 12.7   Problem resolved using pulsed x-ray exposure and progressive scan mode.

its target for more than one readout period. This is especially important where rapidly moving objects, such as the heart, are being imaged by the digital radiographic system. A type of analog video camera that has good linearity and low lag is the plumbicon target camera. The plumbicon is a videcon with a lead oxide (PbO_2) target. Another benefit of plumbicon cameras is that they have excellent electronic noise characteristics in comparison to other types of video cameras, with a dynamic range of 1000:1 and perhaps as high as 2000:1. Newer digital cameras can have high dynamic range with minimal lag with more pixels.

Commercial television frame size has increased to 1280 × 720 (horizontal × vertical) or 1920 × 1080, and these are referred to as 720 or 1080 based on their vertical size. The 1920 × 1080 size is called 2K due to its nearly 2000 horizontal specification. There are also versions of each that are interlaced (1080i) or progressive scan (720p). Frame rates are similar to earlier standards with interlaced scans of 1/30 second per frame and 1/60 second per field. Another standard that is becoming popular has 4096 × 2160 pixels and is called 4K, though its spatial resolution is well beyond the needs of DSA.

12.4 DIGITAL IMAGE PROCESSOR

The block diagram and component descriptions for a typical digital image processing system for DSA are shown in Figures 12.8 and 12.9. The basic functions include (1) acquiring and digitizing the video images, (2) storing the digital images in random access memory, (3) performing arithmetic operations on the images (subtraction, addition, and multiplication), (4) displaying images on monitors, and (5) storing the images as files on various storage media. The image processor contains a microprocessor or system controller that manages the basic operation of the x-ray generator, and other analog components, as well as coordinating and controlling operations of the digital imaging system (Figure 12.1).

12.4.1 Analog-to-Digital Conversion

We will assume that an analog video image has been acquired by the x-ray system, image intensifier, and video camera (Figure 12.8). The analog signal is preprocessed to adjust the amplitude, level, and bandwidth of the video signal to match the signal range and sampling rate of the ADC. The step size of the ADC is selected so that it does not introduce additional noise to the image signal after digitization, which is approximately equal to the standard deviation of the electronic noise. Because the SNR of most video cameras is ~1000:1, the dynamic range of the video signal must be covered by more than 1000 quantization steps, corresponding to an ADC with 10 bits (2^{10} = 1024 steps) or more. The temporal sampling rate (samples/time) of the ADC is selected to capture the desired number of pixels

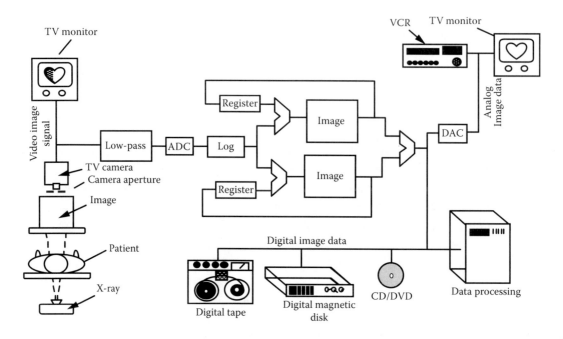

FIGURE 12.8   Hybrid DSA x-ray system. This is a graphic version of Figure 12.1.

Image intensifier
  CsI input phosphor: Excellent x-ray absorption
  Titanium window: High contrast ratio (low light scattering)
  Spatial resolution: Can be relaxed for digital imaging

Variable light aperture
  Small aperture: Greater patient exposure for improving SNR
  Large aperture: Lower patient exposure at lower SNR

Video camera
  Camera type: Vidicon with lead oxide (PbO_2) target (Plumbicon)
  Camera response: Output signal proportional to light input signal
  Temporal characteristics: Low lag
  Signal-to-noise ratio: 1000:1 to 2000:1
  Scanning modes: Progressive scanning for best dose utilization
            Interlaced scanning used in some cases

FIGURE 12.9   Hybrid DSA system's components.

per image frame. The location of a sample within a scan line determines its column index (zero starts on left), and the scan line number determines its row index (zero starts at top of video image). The controlling microprocessor formats these data into an image matrix where sampled values at each location can be indexed using a row, column scheme. Each location in the digital image is called a picture element or "pixel" and the associated value is called the pixel value. Time is recorded for each image in a series acquisition.

Angiographic images may be acquired in a 512×512-pixel matrix though some systems use a 1024×1024 or larger matrix. The image matrix and image-framing rate determine the temporal sampling rate of the ADC. For example, if a 512×512 image matrix is used to digitize an image that is acquired over 1/30 of a second, the sample period of the ADC equals 1/30th second (time per image) divided by 512^2 (number of pixels per image). The sampling period is approximately 100 ns/pixel, corresponding to a sampling rate of about 10 MHz. This sampling rate limits the bandwidth of the digital system to approximately 5 MHz and preprocessing of the analog signal must be done using a low-pass filter to avoid aliasing by limiting the spatial frequency of the incoming analog video signal to 5 MHz or less. In newer DSA systems, the functions of the analog video camera, ADC, and framing microprocessor are combined in a digital video camera.

12.4.2 Logarithmic Transformation

Following digitization, image data are logarithmically transformed, meaning that the pixel values are replaced by their logarithm. The logarithmic transformation is required to remove stationary anatomical structure during image subtraction (Section 12.4.4). The logarithmic transformation can be done on the analog signal prior to digitization with an analog logarithmic amplifier (i.e., a specialized operational amplifier). However, most imaging systems currently perform the logarithmic transformation following the ADC with a digital look-up table that simply replaces each digital value with a new value proportional to its logarithm.

12.4.3 Image Memory and Integration Feedback Loop

After digitization and logarithmic transformation of the incoming video signals, images are stored in the memory of the image processor. Often more than one image from a series is summed (integrated) to reduce noise. This averaging can be done on the fly where incoming images are added to previously stored images. If the image processor uses a 10-bit ADC, image memories must have enough bits to deal with this. Many 10-bit images (max value= 1,023) can be averaged or added to reduce the noise without overflowing the maximum value for systems using 16-bit memory (max value = 65,535). A signed 16-bit format is used to support negative values (ranging from −327,678 to 32,667) with half the maximum value. Virtually all DSA image processors have more than one memory bank. This requirement is obvious in the case of DSA where a mask image is acquired in one memory and then subtracted from an opacification image acquired in a second memory. Where a series of images are acquired, and with sufficient memory, all can be stored in digital memory in the image processor. However, this is expensive, so it is advantageous to process on the fly and store the subtraction images on high-speed digital storage media. Many systems also store unprocessed images, which are later processed to test different approaches.

12.4.4 Image Subtraction

In DSA, at least two images are acquired. The first is called the "mask" image, which is obtained before contrast media is injected into the patient. The second is called the "contrast" or "opacification" image, obtained following injection of the contrast media when the contrast bolus reaches the artery to be imaged.

The signals over the artery in mask and opacification images can be modeled mathematically by assuming that the patient has a thickness x_t and a linear attenuation coefficient of μ_t.

Before contrast media is injected into the patient, the photon fluence through the artery and input to the image intensifier is

$$I_m = I_0 e^{-\mu_t x_t}. \tag{12.1}$$

The contrast media is then injected to opacify the artery. If the artery has a thickness x_I (where $x_I \ll x_t$) and has a linear attenuation of μ_I, the photon fluence is reduced as follows:

$$I_I = I_0 e^{-(\mu_t x_t + \mu_I x_I)}. \tag{12.2}$$

To simplify this example, we assume that x_t is unchanged. If α is the conversion factor, which scales the video signal to the photon fluence received by the image intensifier, the mask and opacification image signals produced by the video camera are

$$I_m = \alpha I_0 e^{-\mu_t x_t}, \tag{12.3}$$

$$I_I = \alpha I_0 e^{-(\mu_t x_t + \mu_I x_I)}. \tag{12.4}$$

We now will use Equations 12.3 and 12.4 to demonstrate the difference between subtraction of the images without logarithmic transformation (linear subtraction) and subtraction of the images following logarithmic transformation (logarithmic subtraction).

12.4.5 Linear Subtraction

Some of the early investigators of digital subtraction techniques used linear subtraction in an attempt to isolate the opacification signal. In linear subtraction, the opacification image is subtracted from the mask image without logarithmic transformation. Linear subtraction produces a subtraction image (S_{lin}) having the following form:

$$S_{lin} = I_m - I_I = \alpha I_0 e^{-(\mu_t x_t)} - \alpha I_0 e^{-(\mu_t x_t + \mu_I x_I)} = \alpha I_0 e^{-(\mu_t x_t)}[1 - e^{-(\mu_I x_I)}]. \tag{12.5}$$

If we assume a small iodination signal such that $\mu_I x_I \ll 1$, then $e^{-\mu_I x_I} \approx 1 - \mu_I x_I$ and

$$S_{lin} = \mu_I x_I \left(\alpha I_0 e^{-(\mu_t x_t)} \right) = \mu_I x_I \left(I_m \right). \tag{12.6}$$

This shows that, when using linear subtraction, the opacified signal ($\mu_I x_I$) is modulated by the patient mask image I_m, that is, patient anatomy. Therefore, linear subtraction produces images that retain the unwanted patient anatomy superimposed on the desired opacified arterial image.

12.4.6 Logarithmic Subtraction

In comparison to linear subtraction, logarithmic subtraction does not retain stationary anatomical structure that can obscure the small signal contributed by the opacified artery. The mask and opacification images are subtracted after they are logarithmically transformed. Mathematically, the logarithmic subtraction image $S_{\log}$ is

$$S_{\log} = \ln(I_m) - \ln(I_I) = [\ln(\alpha I_0) - \mu_t x_t] - [\ln(\alpha I_0) - \mu_t x_t - \mu_I x_I] = \mu_I x_I. \qquad (12.7)$$

Therefore, for logarithmic subtraction, the opacification signal ($\mu_I x_I$) is not modulated by the patient thickness or the anatomy on which the opacified artery is superimposed. However, the resulting signal-to-noise level will be low and must be managed.

12.4.7 Image Display and Archival

Digital images are delivered to a digital-to-analog converter that produces standard video signals that can be viewed on a video monitor. The analog video signal can be stored on analog or digital video media to share with others. The dynamic range of older technology (i.e., analog tape or disk recorders) was approximately 200:1, in comparison to the 1000:1 dynamic range of video cameras, which we previously discussed. Conversion to digital video format for storage helps to preserve the dynamic range of the digital radiographic images.

However, digital image storage is the preferred method for absolute fidelity of the image data. This includes storage on magnetic disk, magnetic tape, optical disk, and digital memory devices (card or plug-in). Digital storage keeps the image data in its original form and virtually avoids the possibility of noise being added by conversion to video and by the storage media. Digital image storage can be done before subtraction and is often used when extensive processing (image integration or iodine quantification) is needed.

12.5 NOISE IN DIGITAL SUBTRACTION ANGIOGRAPHY

In Chapter 8, we discussed how quantum statistical noise σ_q, electronic noise σ_e, and digital quantization noise σ_Δ all contribute to system noise, and that if we design our DSA system correctly, quantization noise will be negligible.

By way of summary, if a radiographic signal is composed of N photons per pixel, then the uncertainty (i.e., standard deviation) in that signal is $N^{1/2}$, since photon generation and attenuation behave according to Poisson statistics. Assuming that the system responds linearly to the input (Figure 12.10), it would produce a maximum video output of V_{max} at N_{max} photons per pixel. Based on this relationship, a pixel's video signal V_N corresponding to N photons is given by the proportionality relationship:

$$\frac{V_N}{V_{max}} = \frac{N}{N_{max}} \quad \text{or} \quad V_N = \frac{V_{max}}{N_{max}} N, \qquad (12.8)$$

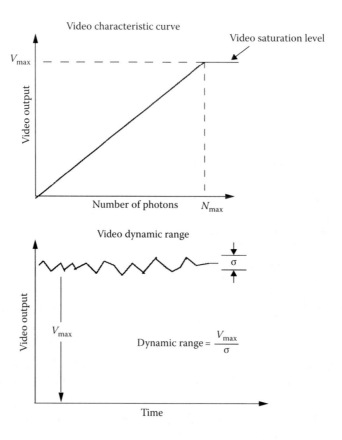

FIGURE 12.10   In an ideal video camera, the video signal output is proportional to the level of light exposure up to a maximum saturation level (V_{max}). The video dynamic range is defined as the maximum video signal divided by the standard deviation of the noise (σ) in the video signal.

where both N_{max} and V_{max} are constants for a system's configuration. The uncertainty in the video signal per pixel due to quantum statistical sources (σ_q) scales like the signal and is

$$\sigma_q = \left(\frac{V_{max}}{N_{max}}\right)\sigma_N = \left(\frac{V_{max}}{N_{max}}\right)\sqrt{N}. \qquad (12.9)$$

The electronic noise contributed by the video camera is typically characterized in terms of the camera's dynamic range, defined to be the ratio of the peak video signal V_{max} divided by the standard deviation (σ_e) of the video camera (Figure 12.10). If D is the dynamic range of a video camera, then the standard deviation of the electronic noise from the camera σ_e is given by

$$\sigma_e = \frac{V_{max}}{D}. \qquad (12.10)$$

Finally, the signal quantization error or quantization noise is the error introduced into the signal when the pixel value is digitized from the analog signal. If Δ is the width of the quantization step in volts (i.e., the interval associated with the least significant bit of the ADC) where all analog values from $\mu_I - \Delta/2$ to $\mu_I + \Delta/2$ are equally likely and are converted into the value μ_I by the ADC, then the variance of the signal quantization error is

$$\sigma_\Delta^2 = \frac{\Delta^2}{12}. \tag{12.11}$$

Note that all three of these noise components are measured using the same units (volts).

12.5.1 System Noise in DSA

The system noise variance in DSA is obtained by adding the noise variance from each noise component; assuming that these noise contributions are independent. For this calculation, we assume that the imaging system consists of a video camera viewing the output phosphor of the image intensifier, and that the quantization error/noise contributed by the ADC is negligible.

The camera output contains a component V_q proportional to the exposure to the input phosphor as well as a time-varying term V_e arising from the "dark current" of the system.

$$V = V_q + V_e. \tag{12.12}$$

The uncertainty in the video output can be calculated using propagation of errors

$$\sigma_v^2 = \sigma_q^2 + \sigma_e^2, \tag{12.13}$$

where contributions from the video dark current σ_e and quantum statistical sources σ_q add in quadrature to give the total noise σ_v in the video system.

Using the expressions derived (Equations 12.9, 12.10, and 12.13), we can obtain the noise in the video signal (σ_v) in terms of the number of photons (N), that is, number of photons per pixel, the dynamic range (D) of the video camera, the maximum video level (V_{max}), and the number of photons per pixel (N_{max}) corresponding to the maximum video level as follows:

$$\sigma_v^2 = \left[\frac{V_{max}}{N_{max}}\right]^2 N + \left[\frac{V_{max}}{D}\right]^2. \tag{12.14}$$

From Equation 12.14, and with signal V, the SNR is

$$SNR = \frac{V}{\sigma_V} = \frac{V}{\sqrt{\dfrac{V_{max}^2}{N_{max}^2} N + \dfrac{V_{max}^2}{D^2}}} = \frac{V/V_{max}}{\sqrt{\dfrac{N}{N_{max}^2} + \dfrac{1}{D^2}}}. \tag{12.15}$$

Since the analog signal is proportional to the number of photons (N) per pixel in the digital image,

$$\frac{V}{V_{max}} = \frac{N}{N_{max}},$$ (12.16)

we can express the analog SNR in terms of the number of photons (per pixel) N absorbed at the input phosphor of the image intensifier as

$$\text{SNR} = \frac{N/N_{max}}{\sqrt{\dfrac{N}{N_{max}^2} + \dfrac{1}{D^2}}} = \frac{N}{\sqrt{N + \dfrac{N_{max}^2}{D^2}}},$$ (12.17)

We will investigate the SNR given in Equation 12.17 in several important special cases.

12.5.2 Special Cases

Case 1. High dynamic range (ideal system):

If the dynamic range D is large compared to N_{max}, then the system SNR (Equation 12.17) reduces to

$$\text{SNR} = \sqrt{N}$$ (12.18)

such that electronic noise is negligible and the DQE of the system is unity. Thus, when the video dynamic range is very large, the noise is contributed entirely by photon statistics and is Poisson-distributed. This case is not realistic given that N_{max} should be large for imaging and that D is limited to a range of 1000–2000.

Case 2. High signal levels ($N = N_{max}$, Figure 12.11, top graph):

If we are operating at a high signal level where the number of photons used to generate the image is at the maximum value for the system, then

$$\text{SNR} = \frac{N_{max}}{\sqrt{N_{max} + \dfrac{N_{max}^2}{D^2}}} = \frac{1}{\sqrt{\dfrac{1}{N_{max}} + \dfrac{1}{D^2}}}.$$ (12.19)

For example assume an exposure of 16 mR, 0.5 mm × 0.5 mm pixel area, energy = 60 keV, and dynamic range $D = 1000$. At 60 keV, the mass energy absorption coefficient of air is

$$\left(\frac{\mu_{en}}{\rho}\right)_{air} = 0.0289 \text{ cm}^2/\text{g}$$ (12.20)

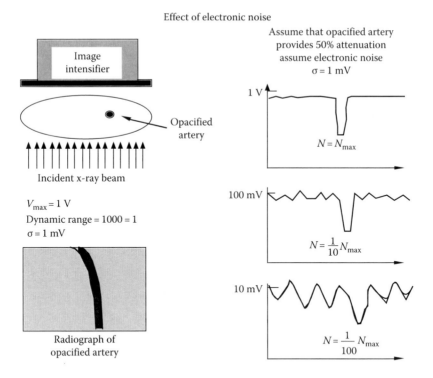

FIGURE 12.11 The best SNR in a subtraction image is obtained when the image data are acquired at the highest possible video signal level. Because video noise is approximately constant in amplitude, the SNR diminishes as the video signal decreases.

so that the photon fluence at 60 keV corresponding to an exposure of $X = 16$ mR is (using Equation 8.17)

$$\Phi = 5.03 \times 10^8 \text{ photons/cm}^2. \tag{12.21}$$

Over a 0.5 mm × 0.5 mm pixel area, the number of photons is

$$N_{max} = (5.03 \times 10^8 \text{ photons/cm}^2) (0.05^2 \text{ cm}^2/\text{pixel}) = 1.26 \times 10^6 \text{ photons/pixel}. \tag{12.22}$$

Therefore, the system's SNR (i.e., the ratio of the maximum signal V_{max} divided by the standard deviation of the system) is

$$\text{SNR} = \frac{1}{\sqrt{\dfrac{1}{N_{max}} + \dfrac{1}{D^2}}} = \frac{1}{\sqrt{\dfrac{1}{1.26 \times 10^6} + \dfrac{1}{1 \times 10^6}}} = 747. \tag{12.23}$$

At this point, we ask whether this signal-to-noise level is adequate to see a small low-contrast object in a noisy image. This question can be answered approximately using the Rose model that relates SNR, contrast, size and fluence. The Rose model equation is

$$k^2 = C^2 \Phi A = C^2 N, \qquad (12.24)$$

where
    $k = 5$ (a constant specifying the SNR at which the low contrast signal is visible)
    $N = \Phi A$ is the number of photons used to image object of area A
    C is the contrast level of signal $= \Delta N/N$
    Φ is the photon fluence in background region $= N/A$
    A is the area of object
    ΔN is the difference in number of photons used to image the object and a background
        region of equal area

Therefore, from the Rose model,

$$k^2 N = C^2 N^2 = \left(\frac{\Delta N}{N}\right)^2 N^2 = (\Delta N)^2. \qquad (12.25)$$

Let $N = \sigma^2$ be the variance of the noise and taking the square root of the leftmost and rightmost terms in Equation 12.25 leads to an alternative way to express the Rose model:

$$k\sigma = \Delta N. \qquad (12.26)$$

Intuitively, this states that the difference in the background and object's signals must be k times the standard deviation of the noise. From our derived value of the SNR at 100% contrast,

$$\text{SNR} = N/\sigma \qquad (12.27)$$

and rearranging to follow the format of (12.26), we have

$$\text{SNR}\ \sigma = N. \qquad (12.28)$$

Now dividing Equation 12.26 by Equation 12.28 and substituting values for k from the Rose model equation and SNR from Equation 12.23 yields

$$C = \Delta N/N = k/\text{SNR} = 5/747 = 0.67\%. \qquad (12.29)$$

Thus, the operating level in this example where the image is at its maximum exposure level, the minimal perceptible contrast is

$$C = \Delta N/N \approx 1\%. \qquad (12.30)$$

Note that this contrast limit is approximately defined as the ratio of the SNR at the threshold of detectability ($k = 5$) to the SNR produced by this imaging system where $N = N_{max} = 747$.

Case 3. If $N = 1/10 N_{max}$, Figure 12.11, middle graph, then

$$\text{SNR} = \frac{N}{\sqrt{N + \dfrac{N_{max}^2}{D^2}}} = \frac{\dfrac{N_{max}}{10}}{\sqrt{\dfrac{N_{max}}{10} + \dfrac{N_{max}^2}{D^2}}}. \tag{12.31}$$

With $N_{max} = 1.26 \times 10^6$ photons and $D = 1000$ as before, we have SNR = 96.3. From Equation 12.29, when SNR = 96.3, an observer would be able to see a minimal contrast level of

$$C = \Delta N/N = k/\text{SNR} = 5/96.3 = 5.2\%. \tag{12.32}$$

Case 4. If $N = 1/100 \, N_{max}$, Figure 12.11, bottom graph, then

$$\text{SNR} = \frac{N}{\sqrt{N + \dfrac{N_{max}^2}{D^2}}} = \frac{\dfrac{N_{max}}{100}}{\sqrt{\dfrac{N_{max}}{100} + \dfrac{N_{max}^2}{D^2}}}. \tag{12.33}$$

With $N_{max} = 1.26 \times 10^6$ photons and $D = 1000$, we have an SNR = 9.96 in which case only contrast levels greater than 50% are visibly detectable in the image.

$$C = \Delta N/N = k/\text{SNR} = 5/9.96 = 50.2\%. \tag{12.34}$$

At this point, the SNR of the image is severely limited by the electronic noise, not by the quantum statistics. It is important to point out that Cases 2 through 4 exist simultaneously within an image, so we need to use care in applying these when we are assessing the minimally detectable contrast.

12.6 METHODS TO MANAGE SNR IN DSA

12.6.1 Bright Spots

From the examples discussed, it is obvious that digital radiographs obtained with image intensifier systems will be severely limited by electronic noise if the image is not obtained with the maximum number of photons per pixel N_{max}. What is not obvious is that we often are forced into this undesirable situation since the image contains regions with both high and low x-ray transmission. High transmission is seen at the edge of the patient or in body regions containing air (lungs or bowel gas). If these regions of high x-ray transmission are imaged at maximum video levels, then patient regions of interest with lower transmission will be imaged at lower video levels where the data can be compromised by electronic noise

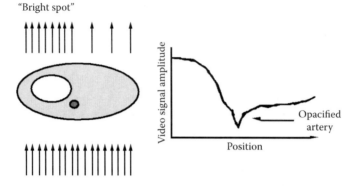

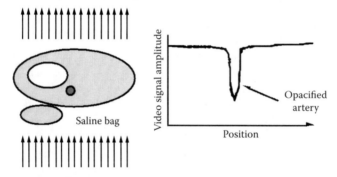

FIGURE 12.12   A saline bag can be used to minimize the effect of bright spots in DSA studies.

(signals closer to electronic noise level). A technique to reduce this problem is to place bags of saline over the high transmission areas "bright areas" or to place pieces of aluminum in the x-ray beam to decrease the exposure to regions where the patient is highly transmissive (Figure 12.12). Ideally, the input exposure field could be a nonuniform exposure field tailored to the patient so that the exposure field reaching the image intensifier in the image region of interest is high, leading to an improved SNR.

12.6.2 Video Camera Aperture

Another method to improve the SNR in DSA is to increase the exposure delivered to the patient, decreasing the noise contribution from quantum statistical sources. However, because a specific light level delivered to the camera target will produce a maximum video response, the x-ray exposure cannot be increased indefinitely without making other adjustments in the system to ensure that this maximum light level is not exceeded in regions of importance. As such, the video camera aperture assumes a fundamental role to control the level of quantum noise in the digitally subtracted angiogram. Because the aperture is located between the output phosphor of the image intensifier and the input optics of the video camera (Figure 12.8), decreasing the aperture diameter also decreases the amount of light reaching the camera target and lowers the camera response for a given x-ray exposure. Correspondingly, the x-ray exposure level must be increased if the aperture diameter is decreased to maintain a constant video signal level. When the camera aperture is decreased, more x-ray photons are used to acquire the image at the quantum sink (i.e., the input phosphor of the image intensifier). Therefore, the overall SNR of the video signal increases (*assuming that the x-ray exposure is adjusted to maintain a maximal video signal in the patient image*). This reduces quantum statistical noise and improves the overall noise characteristics of the image.

It is important to stress the complementary roles of the x-ray exposure level and the video camera aperture. If the x-ray exposure level is increased without adjusting the aperture, then the increased light output of the image intensifier can drive the video camera into saturation, producing a useless signal that saturates at its maximum level. Similarly, decreasing the camera aperture, while holding x-ray exposure fixed, will reduce the light delivered to the video camera, resulting in a smaller video signal. The quantum noise component scales with the video signal, but the SNR of the video signal will be reduced due to the fixed level of electronic noise in the video system. Thus, the camera aperture must be adjusted to provide a video signal near the maximum level to minimize the noise contribution from the electronic signal, and this should be adjusted for the region of interest in the body. However, the x-ray intensity must be limited such that the signal from the image intensifier does not saturate for the region in the image of diagnostic interest.

12.6.3 Image Integration

Another way to improve system SNR is to use various processing schemes that add (or "integrate") images together either before or after subtraction to reduce the noise in the digital radiographic images. The simplest way this can be performed is by means of frame integration where two or more sequential image frames are added together. If M sequential frames are added together, where all the frames are nearly identical except for the random noise content, and if N represents the signal (and $N^{1/2}$ the noise) in each image, then the integrated signal will be MN and the noise $(MN)^{1/2}$. The SNR before image integration was $N^{1/2}$ and that for the integrated frames is $(MN)^{1/2}$, so the SNR increases by $(MN)^{1/2}/N^{1/2} = M^{1/2}$. *Frame integration has the advantage of reducing both the effects of quantum noise and electronic noise sources.* By comparison increasing radiation exposure per frame only reduces quantum noise effects. However, frame integration has the serious disadvantage that it is more prone to motion blurring, since each integrated image is an average over a longer time period than the original frames.

12.7 SPATIAL RESOLUTION IN DSA

There are several factors to consider concerning spatial resolution in DSA. The first is the digital matrix size (such as 512×512 or 1024×1024) used to acquire the image data. The second is the spatial resolution of the image intensifier. The third is the degree of geometric unsharpness due to focal spot size. There is a fundamental trade-off between the increase in object detail that can be seen due to image magnification and the loss in object detail due to increased geometric unsharpness. Object detail increases with greater magnification due to the fixed resolution of the image intensifier and digital image matrix. On the other hand, increasing geometric unsharpness (magnification of focal spot) degrades spatial resolution with greater object magnification. The relative effects of these spatial resolution components can be evaluated using a cut-off frequency approach.

If an image is recorded with an object magnification of M and if the focal spot has width a, the width of the focal spot when projected onto the detector is $(M-1)a$. This is equal to a structure in the object plane having a width of $\dfrac{(M-1)a}{M}$. We can define the cut-off frequency u_s due to geometric unsharpness in the object plane using the inverse of this width as

$$u_s = \frac{M}{(M-1)a}. \tag{12.35}$$

(see Figure 12.13a)

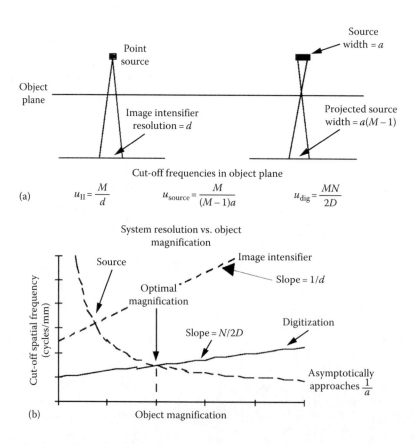

(a)

(b)

FIGURE 12.13 (a) The spatial resolution of a DSA image can be limited by geometric unsharpness from the x-ray focal spot, by image intensifier resolution, or by the digital image matrix. (b) At low magnification, resolution is limited by the image matrix, but at a higher magnification, geometric unsharpness limits the resolution.

The derivation of the cut-off frequency based on width is as follows: (1) we model the projected spread function as a rectangle, (2) the Fourier transform of a rectangle is a sinc function, and (3) the value of *the sinc function first goes to zero at a frequency equal to 1/width, which is called the cut-off frequency.*

Similarly, the resolution width of d for the image intensifier is d/M in the object plane, giving a cut-off frequency for the image intensifier of

$$u_{\text{II}} = \left(\frac{1}{d}\right)M. \tag{12.36}$$

The cut-off frequency due to the digitizer is based on the Nyquist frequency limit, since no frequencies above that limit can be displayed. If the width of the field of view (FOV) of the detector is D and the image is digitized into an $N \times N$ matrix with object magnification of M, the limiting spatial frequency u_{dig} imposed by the digitization process is

$$u_{\text{dig}} = \frac{N/2}{D/M} = \left(\frac{N/2}{D}\right)M. \tag{12.37}$$

Here, $N/2$ is the limiting frequency for the FOV (lp/FOV) and D/M is the width of the FOV at the object plane (mm/FOV).

The trade-off between the decrease in detector unsharpness and the increase in geometric unsharpness with increasing object magnification is best seen by graphing the cut-off frequencies for focal-spot blurring and for detector response as a function of object magnification. As shown in Figure 12.13b, the curve labeled "source" shows how increasing magnification produces the resolution loss (i.e., decreasing the cut-off spatial frequency) due to geometric unsharpness and increased resolution for the image intensifier and digitization.

Example 12.1

A focal spot size of approximately 1 mm is common in DSA, and according to Equation 12.35, this corresponds to a cut-off frequency of

$$u_{\text{s}} = \frac{M}{M-1} \text{ lp/mm.} \tag{12.38}$$

We will consider two different components of detector resolution, one from the image intensifier and the other from the digital image matrix. The image intensifier (II) has a spatial resolution of about $d = 0.2$ mm, and according to Equation 12.36, the cut-off frequency is

$$u_{\text{II}} = 5M \text{ lp/mm.} \tag{12.39}$$

If we assume that a 512×512 image matrix is used to digitize an inscribed circular FOV with a diameter of 23 cm (approximately 9 in.), then according to Equation 12.37, the digital cutoff frequency is

$$u_{dig} = \left(\frac{N/2}{D}\right)M = \left(\frac{256\dfrac{lp}{FOV}}{230\dfrac{mm}{FOV}}\right)M = 1.11M\frac{lp}{mm}. \qquad (12.40)$$

In this example, the digital image matrix rather than the image intensifier limits the spatial resolution of the detector at low magnification. However, at higher magnification, geometric unsharpness becomes more problematic. For this example, the intersection of the cut-off spatial frequency curve for the digital image matrix with that for focal spot blurring indicates the magnification at which the cut-off frequency would be highest, yielding the best spatial resolution. Therefore, the optimal magnification for this system would be $2.11/1.11 \sim 1.9$.

HOMEWORK PROBLEMS

P12.1 You are visiting a hospital that has just acquired a DSA system. You watch as the radiologist sets the kVp and mA to obtain an exposure level that gives a maximum video signal output. The radiologist then looks at the settings on the x-ray console and decides to decrease the patient exposure and tolerate the accompanying increase in noise by decreasing the x-ray tube current (mA) to 25% of its original value. You immediately remember that it is important to adjust the video camera aperture for this new exposure level and alert the radiologist of the problem. The radiologist thanks you for the information, readjusts the aperture to achieve a maximum video signal output with the lower exposure level, and then continues the examination.

For the following calculations, assume that the dynamic range of the video camera is 1000:1. Before the radiologist decreased the exposure or increased the camera aperture, a maximum video signal was obtained at a setting of 70 kVp and 200 mA. At this setting for a 40 ms exposure, the x-ray tube produces a photon fluence of 1.031×10^6 photons over a resolution area of 1 mm².

(a) When the initial exposure settings were established, the diameter of the aperture was 4 mm, corresponding to an f-stop of f/5.6. What is the correct diameter and f-stop of the camera aperture after the patient exposure has been decreased to 25% of its original value?

(b) How is the SNR of the image affected if the patient exposure is reduced by a factor of 4, but the camera aperture is not adjusted for the new exposure level?

(c) What is the SNR of the image if the patient exposure is reduced by a factor of 4 and the camera aperture is adjusted to maintain a maximal video signal output?

(d) Explain to the radiologist why the video camera aperture must be adjusted for the new exposure level. Use your calculations from parts (b) and (c) of this problem to guide your thought processes, but give the radiologist conceptual and intuitive (rather than quantitative) explanations.

P12.2 You take a research position in an x-ray imaging laboratory at a major university in the United States. Your first development is a new video camera especially designed for pediatric coronary angiography. Because the infant heart beats at a higher rate than that of the adult, the camera has a video frame rate of 100 fps. Second, because you want to limit both the radiation exposure and the amount of contrast media required for the angiographic study, the camera is designed with a dynamic range of 4000 to 1. Finally, because the infant is small, the camera is designed with a video signal bandwidth high enough to be compatible with a 2048 × 2048 digital image matrix. You then begin to develop a digital image processor that is compatible with this new video camera.

All of a sudden, your grandmother drops by with a batch of chocolate chip cookies. Grandma has been very lonely lately since grandpa bought a new Macintosh computer and discovered "Internet browsing," so she starts asking lots and lots of questions. While munching on her delicious cookies, you try to explain to grandma what you are doing.

(a) Grandma first asks you what a "video signal bandwidth" is, and why a higher "video signal bandwidth" is needed if you want to obtain a 2048 × 2048 (rather than a 512 × 512) image. What problems will you encounter if you have too low of a bandwidth? What problems will you encounter if your camera has too high of a bandwidth? (By the way, if you use the word "aliasing" in her presence, Grandma will make you wash your mouth out with soap.)

(b) Grandma then asks you what a "video dynamic range" is and why a higher "video dynamic range" might let you decrease the radiation exposure and amount of contrast media needed for the study.

As you are explaining these things to grandma, a doctor who is walking through the corridor outside the door of your laboratory distracts her. Feeling rather frisky because of the lack of attention by grandpa, grandma decides to chase after this new target rather than listen to you talk about medical imaging. As she runs out the door, she almost knocks over Bruce Hasegawa who overheard your conversation; he has lost interest in a nurse he was talking to in the hall, and has entered the room to join the conversation. Noting the expression of intense curiosity on his face, you moan, wishing grandma were asking questions rather than Bruce. By now, grandma has left you alone with Bruce who, to your horror, is busily gobbling up the cookies and who also is starting to ask some questions about your new invention.

(c) How many bits are needed by the ADC to minimize quantization errors in the digitized image data?

(d) If you eventually want a 2048 × 2048 image at a rate of 100 video frames per second, what is the digitization rate of the ADC for a digital image processor compatible with this video camera? What is the maximum video bandwidth the camera should deliver to be consistent with the sampling rate of the ADC?

(e) The image intensifier has two different operating modes, one with a 6 in. diameter FOV and other with a 9 in. diameter FOV. What spatial frequency (in cycles/mm), as measured in the detector plane, will the 2048 × 2048 image matrix support for each of the two operating modes?

(f) Assume that your image intensifier has a point-spread function width of 200 microns and that you operate the image intensifier with a 6 in. diameter FOV with an x-ray tube having a 300-micron focal spot size. Determine the optimal magnification for imaging the infant heart and the corresponding optimal cut-off spatial frequency for this system? Is this magnification compatible for imaging a 2 in. diameter infant heart and sufficient to see 400-micron-diameter coronary arteries?

(g) The infant thickness can be modeled as 15 cm of water and the effective energy of the x-ray beam is 30 keV. What should the entrance exposure to the infant be if you want to see a 1% contrast level in a 400-micron-diameter infant coronary artery with your video camera?

(h) Bruce comments that your calculations for parts (c) and (d) of this question show that you need an ADC that is technically impossible to design and build at this time. However, he tells you that the important issue in coronary imaging is one of limiting motion unsharpness rather than one of frame rate. He would rather have a few images each with minimal amounts of motion unsharpness rather than many images acquired over the entire cardiac cycle. In other words, he prefers an imaging system with a short acquisition time per frame, but does not need one with a video frame rate faster than about 15 fps.

    With this in mind, you start thinking about using a video camera with a progressive readout to be used in combination with a short x-ray exposure time. If you want to maintain the optimal spatial resolution as defined by your calculation in part *f*, how short of an exposure time do you need if the infant's myocardium has a maximum velocity of 5 cm/s at the point of maximum cardiac contraction.

P12.3 Let us think a little about digital look-up tables such as those utilized in DSA for the logarithmic transformation. (We can offer you a hint that a digital look-up table is built out of random access memory).

(a) How might you design a digital look-up table in hardware? How would you load the digital look-up tables with the transformed values? What circuit design would allow you to deliver digital values and obtain the transformed values from the digital look-up table? How much random access memory (RAM) is needed for a look-up table with a 10-bit input and a 10-bit output?

(b) If the look-up table has a 10-bit input, the largest value that will be delivered to the look-up table will be 1023 (i.e., $2^{10}-1$). However, the natural logarithm of 1023 is 6.93. Describe how we can represent the logarithmically transformed values if we want a 10-bit output from the look-up table. You should be able to give a specific mathematical algorithm to make this possible. What effect will this have on the digitally subtracted angiograms?

P12.4 Compare the contrast resolution (at peak video signal) associated with a TV fluoroscopy system with a dynamic range of 1000:1 for the following cases. Assume that the dynamic range refers to a characteristic resolution element of (0.5 mm^2).

(a) Fluoroscopy at an exposure of 1 mR per image.

(b) Digital radiography at an exposure of 1 R per image (neglect digitization noise)

CHAPTER 13

Temporal Filtering

13.1 BACKGROUND

The primary problem in dynamic imaging such as digital subtraction angiography is the presence of noise in the subtracted images. This situation arises because the opacification signal in these studies occupies only a small portion of the video signal's dynamic range, which is mostly dominated by the patient's anatomical structure. For example, assume that the mask image I_m and opacification image I_o are acquired, each with a signal-to-noise ratio of 1000:1. The opacification image, acquired following arterial injection of the contrast media, can be subtracted from the mask image to determine a positive signal difference image (S):

$$S = I_m - I_o. \tag{13.1}$$

Inspection of this image will reveal that the signal due to opacification in the artery is a small portion of the dynamic range of the video signal. Indeed, we can approximate the arterial signal difference at 1% of the maximum video signal V_{max}. Second, if the noise variances in the mask and opacification images are uncorrelated and both equal to σ^2, then the noise variance in the subtracted image is twice that in the unsubtracted images:

$$\sigma_s^2 = 2\sigma^2. \tag{13.2}$$

Therefore, if the signal-to-noise ratio of the unsubtracted images is

$$\text{SNR} = \frac{V_{max}}{\sigma} = 1000 \tag{13.3}$$

such that $V_{max} = 1000\sigma$, then the signal-to-noise ratio of the arterial signal difference image S is

$$\text{SNR}_s = \frac{\text{Opacification signal}}{\sigma_s} = \frac{0.01 V_{max}}{\sqrt{2}\sigma} = \frac{10\sigma}{\sqrt{2}\sigma} = 7.07. \tag{13.4}$$

Even though the mask and opacification images are acquired at a signal-to-noise ratio of 1000:1, most of the signal is contributed by the patient's anatomy (~99%). As a result, after subtraction, the image has a low signal-to-noise ratio over the artery, <10 in this example. In actual clinical practice, the signal-to-noise ratio of a subtracted angiogram is even lower due to the existence of bright spots or limitations in the electronic imaging system. This calculation and these comments emphasize that digital subtraction angiography is an examination in which the diagnostic task will be limited by noise. The noise-limited nature of the images forces us to consider various methods to increase the signal-to-noise ratio of the resulting subtracted angiograms. *This example was provided without logarithmic subtraction to emphasize the nature of the raw signals.*

As mentioned in the previous chapter, there are several ways in which the signal-to-noise ratio can be improved. The *first* is utilization of an image intensifier with high detection efficiency and contrast ratio, a video camera (or other electronic detector) with the low electronic noise, and an analog-to-digital converter with an adequate number of levels so that it does not introduce quantization errors into the image data. The *second* is the use of bolusing so that bright spots do not compromise the opacification signal in regions of interest. The *third* is adjustment of the camera aperture so that the region of interest is acquired with a maximal video signal but at the smallest radiation exposure consistent with the quantum statistical requirements of the study. The last method is that of "image integration" in which multiple images containing the opacification signal are combined using a weighted averaged to reduce effects of random noise (from both electronic and quantum statistical sources), while preserving the opacification signal. The average is taken over sequential time images and is therefore called "temporal filtering" to distinguish the approach from "spatial filtering."

The basic theory of temporal filtering methods is provided in Sections 13.2 and 13.3. We then will present three different temporal filtering techniques (mask-mode subtraction, matched filtering, and recursive temporal filtering), describe them mathematically, and discuss the advantages and disadvantages of each.

13.2 MATHEMATICAL CONVENTIONS

Each image in a sequence of dynamic images will be designated as I_i where "i" is the image index. This is a mathematical simplification of the 3-D array I(row, col, time) that indexes both position within 2-D images and time within the sequence. For this simplification, we drop the row and col indices (same for all images) and index each 2-D array by time using the subscript "i." The image formed by summing or subtracting such images will be designated as S. For example, in the expression

$$S = I_2 - I_1, \tag{13.5}$$

each pixel in the image S is calculated by subtracting pixel values in I_1 from pixel values in I_2 at corresponding row–column locations at the time indicated by the 1 and 2 indices.

13.3 THEORY

Assume we have a sequence of N images I_i where $i = 1, 2, ..., N$. The images have both a static component, representing the stationary anatomical background, as well as a dynamic component, that in the case of angiography can represent the time-varying arterial opacification contributed by the iodinated contrast media.

The filtering process is represented mathematically in Equation 13.6, where we choose a weight factor h_i for each I_i and sum their product over all time indices to form the single image S:

$$S = \sum_{i=1}^{N} h_i I_i. \tag{13.6}$$

Regions within images without contrast material represent anatomical structures that are not changing over time, so for these images, $I_i = I$. We can remove the constant-signal anatomical structures by choosing the constants h_i such that the sum in Equation 13.6 is zero:

$$S = I \sum_{i=1}^{N} h_i = 0. \tag{13.7}$$

This leads to the constraint that *the sum of h's must be zero to remove constant anatomical structures* as follows:

$$\sum_{i=1}^{N} h_i = 0. \tag{13.8}$$

This equation constrains the sum of the filter weights used in temporal filtering but does not provide individual values for the weights.

For those pixels with a time-varying opacification signal (pixels where contrast is seen), the value of S depends on the relationship between h_i values and the time-course signal due to the iodine bolus as the contrast material moves through the artery. We can calculate the noise variance (σ_s^2) of the filtered image S assuming that each of the images has the same noise variance σ^2:

$$\sigma_s^2 = \sum_{i=1}^{N} h_i^2 \sigma_i^2 = \sigma^2 \sum_{i=1}^{N} h_i^2. \tag{13.9}$$

This equation shows that the variance in S is altered by the sum of squared filter weights used for temporal filtering. The objective of temporal filtering is to choose values for h's that reduce noise and increase the signal S associated with the bolus of contrast material as it traverses an artery. Three approaches for assigning h's will be given.

13.4 MASK-MODE SUBTRACTION

A widely used image-processing technique is *simple integrated mask-mode subtraction* in which the opacification image I_o (average of images post injection of iodine contrast) is subtracted from the mask image I_m (average of images obtained before injection of iodine contrast), producing the subtraction image:

$$S = I_m - I_o. \tag{13.10}$$

For this subtraction, the filtering constants are $h_m = +1$ and $h_o = -1$ and the noise variance calculated from Equation 13.9 is

$$\sigma_s^2 = 2\sigma^2. \tag{13.11}$$

Another mask-mode subtraction technique is called *series mask-mode subtraction* where a single mask image is calculated as the average of images before the bolus injection (0–20 s in Figure 13.1). Then each opacification image is subtracted from the single mask image producing a series of subtraction images. This results in near-zero pixel values for tissues outside of the artery and a bolus signal from the artery after injection (bottom of Figure 13.1).

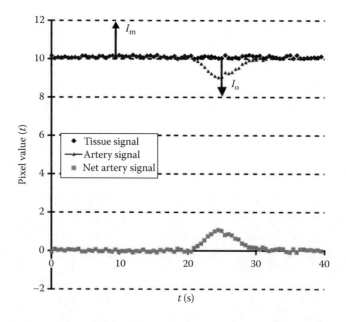

FIGURE 13.1   Pixels values over artery and other tissues before (0–20 s) and after (20–40 s) injection of contrast media (top). Series mask mode subtraction for the artery (bottom).

The resulting image series can be viewed as a subtraction cine where the signal is approximately proportional to the concentration of contrast material. Viewing the subtraction cine allows physicians to evaluate the extent and time course of the contrast material as it passes through the arteries. For simple integrated mask-mode subtraction (Equation 13.10), single mask and opacification images are formed as the sum of preinjection and postinjection images.

Note: All pixel values are the result of some form of scaling and logarithmic conversion. The sense of signals in Figure 13.1 was adjusted such that the contrast media decreased pixel values in the artery.

13.5 MATCHED FILTERING

Both Kruger and Riederer utilized signal-processing theory to propose a "matched filter" to maximize the signal-to-noise ratio in digital subtraction angiographic (DSA) images (Figure 13.2). In this technique, a signal is processed using a filter with a temporal shape, defined by h's, matching the temporal response of the arterial signal. It can be shown that the matched filter is the filter that maximizes the signal-to-noise ratio in the filtered signal. *Matched filtering is a postacquisition processing filtering method.*

We can extend this concept to maximize the signal-to-noise ratio of the opacification signal in DSA images. In particular, we assume that we are imaging an artery through which a contrast bolus represented by a time-varying signal $b(t)$ is passing through but

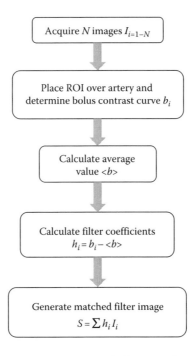

FIGURE 13.2 Matched filter algorithm. In matched filtering, the weighting coefficients are equal to the difference between the bolus signal and its mean value. This is a postprocessing technique, so all images must be acquired before the filter is applied.

which also contains non-time-varying patient anatomy. Regions containing non-time-varying patient anatomy correspond to areas in the image having approximately constant pixel values. Regions containing opacified arteries are represented by pixels containing both a constant anatomical and the superimposed contrast bolus $b(t)$ signals with sequential values represented as b_i.

We will begin by calculating the signal-to-noise ratio in matched filtering and then compare this to the signal-to-noise ratio of mask-mode subtraction. The filtered image is calculated as per Equation 13.6, but here, it is designated as matched filter using a subscript:

$$S_{\mathrm{mf}} = \sum_{i=1}^{N} h_i I_i. \tag{13.12}$$

We choose the filter weights for matched filtering based on the difference between the bolus signal and its average (determined from an ROI placed over the artery)

$$h_i = b_i - \langle b \rangle, \tag{13.13}$$

with positive weights for signals above the bolus average and negative weights for those below. This choice ensures that the sum of weights equals zero, as needed to remove constant anatomical signals. The average of the bolus signal is

$$\langle b \rangle = \frac{1}{N} \sum_{j=1}^{N} b_j. \tag{13.14}$$

In regions of the image where signals are not time varying (i.e., tissues with no contrast media passing through), $I_i = c$, so that

$$S_{\mathrm{mf}} = \sum_{i=1}^{N} h_i I_i = c \sum_{i=1}^{N} h_i = c \left[\sum_{i=1}^{N} b_i - \sum_{i=1}^{N} \langle b \rangle \right] = c \left[N \langle b \rangle - N \langle b \rangle \right] = 0. \tag{13.15}$$

This result shows that the matched filter removes nonarterial areas (the anatomical background) from processed images. In arteries where the signal is time varying as indicated by a bolus $I_i = b_i$, the filter output is

$$S_{\mathrm{mf}} = \sum_{i=1}^{N} h_i I_i = \sum_{i=1}^{N} h_i b_i = \sum_{i=1}^{N} \left[b_i - \langle b \rangle \right] b_i = \sum_{i=1}^{N} b_i^2 - \langle b \rangle \sum_{i=1}^{N} b_i \tag{13.16}$$

$$= N \langle b^2 \rangle - \langle b \rangle N \langle b \rangle = N \left[\langle b^2 \rangle - \langle b \rangle^2 \right] = N \sigma_{\mathrm{b}}^2. \tag{13.17}$$

This result shows that S_{mf} for an artery is equal to N times the variance due to the bolus signal σ_b^2. The image variance (σ_i) is assumed to be constant over time for nonbolus regions such that $\sigma_i = \sigma$, so the variance due to the matched filter (σ_{mf}^2) can be estimated as follows using propagation of variance:

$$\sigma_{mf}^2 = \sum_{i=1}^{N} h_i^2 \sigma_i^2 = \sigma^2 \sum_{i=1}^{N} h_i^2 = \sigma^2 \sum_{i=1}^{N} \left(b_i - \langle b \rangle\right)^2 = \sigma^2 \left(N\sigma_b^2\right). \tag{13.18}$$

The signal-to-noise ratio for the matched filtering technique is therefore

$$\mathrm{SNR}_{mf} = \frac{S_{mf}}{\sigma_{mf}} = \left(\frac{\sigma_b}{\sigma}\right)\sqrt{N}. \tag{13.19}$$

So the SNR increases by $N^{1/2}$ and is proportional to the standard deviation of the bolus and inversely proportional to the temporal standard deviation (noise) of the image sequence. Importantly, if σ_b is larger than σ, the SNR will increase for all values of N. For nonbolus regions where $\sigma_b = \sigma$, the SNR is $N^{1/2}$, the same as for averaging N images.

We will now compute the signal-to-noise ratio for integrated mask-mode subtraction over the artery with bolus signal b_i. We will assume that the first $N/2$ values are integrated to form a mask and remaining $N/2$ are integrated to form the opacification image. The signal is

$$S_{mm} = I_m - I_o = \sum_{i=1}^{\frac{N}{2}} b_i - \sum_{i=\frac{N}{2}+1}^{N} b_i = \frac{N}{2}\left[\langle b_m \rangle - \langle b_o \rangle\right], \tag{13.20}$$

where $\langle b_m \rangle$ and $\langle b_o \rangle$ are averages over each half of the signal range. The noise in this signal is then

$$\sigma_{mm} = \sqrt{N}\sigma \tag{13.21}$$

so that the signal-to-noise ratio is

$$\mathrm{SNR}_{mm} = \left(\frac{\dfrac{\langle b_m \rangle - \langle b_o \rangle}{2}}{\sigma}\right)\sqrt{N}. \tag{13.22}$$

The ratio of the SNRs obtained with matched filtering to that obtained with DSA integrated mask-mode subtraction is

$$\frac{\text{SNR}_{\text{mf}}}{\text{SNR}_{\text{mm}}} = \frac{\dfrac{\sqrt{N}\,\sigma_b}{\sigma}}{\dfrac{\sqrt{N}\left[\langle b_m \rangle - \langle b_o \rangle\right]}{2\sigma}} = \frac{2\sigma_b}{\langle b_m \rangle - \langle b_o \rangle}. \tag{13.23}$$

So whenever σ_b exceeds half of the difference in mean values in the mask and opacified artery, matched filtering increases the SNR beyond that of integrated mask-mode DSA. For an ideal bolus (a rectangle shape where mask and opacification durations are identical with constant levels b_m and b_o), we see that $\sigma_b = (\langle b_m \rangle - \langle b_o \rangle)/2$, so both match filtering and integrated mask-mode subtraction give the same result. However, such an ideal bolus is not seen in practice and matched filtering shows improvement over integrated mask-mode subtraction for realistic boluses. For example, the ratio calculated for the data from Figure 13.1 yielded an SNR improvement of approximately 2X for the matched filter.

Selection of the ROI for determining $b(t)$ for matched filtering.

1. To focus on a particular artery, position the ROI within the artery. Placement will determine the timing of $b(t)$ and will produce best results near the ROI.

2. To improve the noise for the arterial ROI, it can be enlarged slightly along the artery, but can reduce the peak value of $b(t)$ compared with smaller ROI.

3. Let each pixel be an independent ROI. No ROI tracing is needed. This will lead to a large number of $b(t)$ curves (one per pixel), which can enhance all arteries. This is more noisy than methods 1 and 2 but still suppresses anatomy. The entire artery is enhanced using this method.

13.6 MODELING THE BOLUS CURVE

An alternative method to improve SNR is to fit the ROI signal for the range of images covering the changing bolus signal to an analytical model, such as that from 20 to 40 s in the net arterial curve of Figure 13.1. One approach is to use a gamma variate function to model the natural rising and falling characteristics of a bolus curve

$$b(t) = At^B e^{-Ct}, \tag{13.24}$$

where A, B, and C are parameters adjusted to best fit the bolus curve. In practice, $t = 0$ is just prior to the arrival of contrast material. Riederer has suggested the values $B = 4.0$ and $C = 0.9$ s^{-1}, with A chosen such that $b(t_{\max}) = 1$ for a bolus curve with maximum value set to unity. The maximum of $b(t)$ is seen when $db/dt = 0$ so that

$$\frac{db}{dt} = Ae^{-Ct}\left[-Ct^B + Bt^{B-1}\right] = 0, \tag{13.25}$$

showing that the maximum is obtained at

$$t_{\max} = \frac{B}{C} = \frac{4.0}{0.9\,\text{s}^{-1}} = 4.44 \text{ s.} \tag{13.26}$$

Using the values $B = 4.0$, $C = 0.9$ s^{-1}, and $t_{\max} = 4.44$ s, and satisfying the condition that

$$b(t_{\max}) = b(4.44 \text{ s}) = 1 = A(4.44)^4 e^{-0.9*4.44} = 7.146A, \tag{13.27}$$

we find that

$$A = 0.140. \tag{13.28}$$

Thus, a typical contrast bolus curve with t in seconds is given by the following equation:

$$b(t) = 0.140 t^4 e^{-0.9t}. \tag{13.29}$$

In fact, this equation was used to model the data seen in Figure 13.1. Using the range over just the bolus to fit an assumed model is attractive, but may not work as well in nonarterial regions. Fitting of theoretical models of the bolus activity leads to fitted values of A, B, and C. The time point for fitting must be just as the bolus enters the ROI over the artery. The fitted model curve can be used to illustrate the bolus dynamics for various ROIs over the artery.

13.7 RECURSIVE FILTERING

Mask-mode subtraction requires highly stable components including electronic components as well as the x-ray system. Also, this subtraction could not begin until the study data had been acquired. In 1981, Kruger and Gould independently proposed a technique that could accept the signal from a standard fluoroscopy system where filtered images could be acquired continuously, providing real-time subtraction (Figure 13.3). This method, called time-domain recursive filtering, avoids the requirement for timing control of the x-ray system and can be incorporated into a non-DSA fluoroscopy system. In addition, this technique is relatively immune to small instabilities in the video camera or x-ray system, which otherwise might seriously affect images produced by mask-mode subtraction.

The methods of Kruger and Gould both use the recursion relationship

$$S_n = kI_n + (1-k)S_{n-1} \quad n = 1, 2, 3, \ldots \tag{13.30}$$

where
　k is a constant, $0 < k < 1$
　I_n is the nth video frame of an imaging sequence
　S_n is the nth image obtained from recursive filtering process

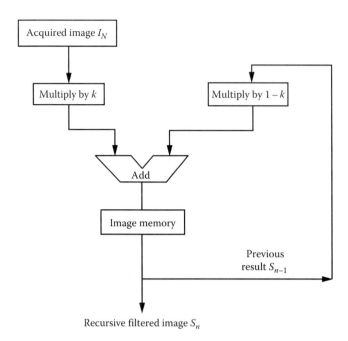

FIGURE 13.3 The recursive filter creates an output image that is a weighted sum of prior images in a time sequence.

If the initial video frame of the imaging sequence is

$$S_0 = kI_0,$$ (13.31)

then the following frames are generated according to Equation 13.30. For example,

$$S_1 = kI_1 + (1-k)S_0 = kI_1 + k(1-k)I_0,$$ (13.32)

$$S_2 = kI_2 + (1-k)S_1 = kI_2 + k(1-k)I_1 + k(1-k)^2 I_0,$$ (13.33)

$$S_3 = kI_3 + (1-k)S_2 = kI_3 + k(1-k)I_2 + k(1-k)^2 I_1 + k(1-k)^3 I_0,$$ (13.34)

and in general

$$S_N = \sum_{m=0}^{N} k(1-k)^m I_{N-m}.$$ (13.35)

We will use this equation to evaluate the response of recursive filtering to a constant signal (stationary anatomy), to transient signals (the iodine contrast bolus), and to random noise.

13.7.1 Response to Constant Signal

If all of the video frames are identical (with the exception of noise) as in the case of stationary patient anatomy, we will show that in the limit of a large number of video frames, the recursive filter returns this constant input. Assume that we have identical frames where

$$I_i = 1 \quad \text{for } i = 1, 2, 3, \ldots, N \tag{13.36}$$

so that the recursive filter generates the new image following the Nth image as

$$S_N = k I_N + k (1-k) I_{N-1} + k (1-k)^2 I_{N-2} + \cdots + K (1-k)^N I_0, \tag{13.37}$$

which, for this special case, equals

$$s_N = k l \left[1 + (1+k) + (1-k)^2 + \cdots + (1-k)^N \right]. \tag{13.38}$$

The geometric sequence in Equation 13.38, with a ratio of $(1-k)$, approaches $1/[1-(1-k)] = 1/k$ for large N such that

$$S_N = I \tag{13.39}$$

so that the recursive filter retains the constant input signal in its output. Therefore, since constant I_i suggests constant patient anatomy, the recursive filter retains that anatomy in the final filtered image, unlike the two previous filtering methods.

13.7.2 Response to Noise

In calculation of response to random noise, we will assume that all images have the same noise variance σ^2. Since the recursive filter generates an image S_N where

$$S_N = k I_N + K (1-k) I_{N-1} + k (1-k)^2 I_{N-2} + \cdots + k (1-k)^N I_0 \tag{13.40}$$

and from propagation of errors we know that

$$\sigma_s^2 = \left(\frac{\partial S_N}{\partial I_N} \right)^2 \sigma_N^2 + \left(\frac{\partial S_N}{\partial I_{N-1}} \right)^2 \sigma_{N-1}^2 + \cdots + \left(\frac{\partial S_N}{\partial I_0} \right)^2 \sigma_0^2, \tag{13.41}$$

$$\sigma_s^2 = k^2 \sigma_N^2 + k^2 (1-k)^2 \sigma_{N-1}^2 + k^2 (1-k)^4 \sigma_{N-2}^2 \ldots k^2 (1-k)^{2N} \sigma_0^2. \tag{13.41a}$$

With all $\sigma_N \sim \sigma$, this becomes

$$\sigma_s^2 = k^2 \sigma^2 \left[1 + (1-k)^2 + (1-k)^4 \ldots + (1-k)^{2N} \right] \tag{13.42}$$

As $N \rightarrow +\infty$, this geometrical sequence with a ratio of $(1 - k)^2$ simplifies to

$$\sigma_s^2 = \frac{k^2 \sigma^2}{1 - (1-k)^2} = \frac{k\sigma^2}{2-k}, \tag{13.43}$$

which becomes infinitesimal for small values of the recursion constant k. Thus, the noise σ_s becomes negligible as k approaches zero. However, inspection of Equation 13.40 shows that the signal also approaches zero under this condition.

For imaging coronary arteries, Kruger has suggested using a value of $k = 1/16$. This means that for large N, the noise level for recursive filtration is

$$\sigma_s^2 = \frac{\frac{1}{16}\sigma^2}{2 - \frac{1}{16}} = \frac{1}{31}\sigma^2 \tag{13.44}$$

so that the standard deviation of the noise in the recursively filtered image is reduced to approximately 18% that in the unfiltered images. This gain is made at the expense of higher patient exposure since in theory, it is obtained only after an infinite number of images have been filtered and combined by the recursive algorithm. In actual use, and as we will see in the following section, the recursive filter has a limited temporal response so that only a finite number of images need to be combined.

13.7.3 Response to Transient Signals

One of the more interesting properties of the recursive filter is its response to a transient signal such as the passage of an iodine contrast bolus through an artery. The output of a recursive filter for a transient (i.e., time-varying) input signal can be approximated as the convolution of the time series with an exponential function. If the output of the recursive filter is

$$S_N = kI_N + k(1-k)I_{N-1} + k(1-k)^2 I_{N-1} + \cdots + k(1-k)^N I_0, \tag{13.45}$$

it can be expressed as weighted sums of individual image terms as $S_N = s_0 + s_1 + s_2 + \cdots + s_N$, where s_m is the mth term of the recursive filter output:

$$s_m = k(1-k)^m I_{N-m}. \tag{13.46}$$

Equation 13.46 can be expressed as

$$s_m = k\left[(1-k)^{\frac{1}{k}}\right]^{mk} I_{N-m} \tag{13.47}$$

And for small values of k,

$$(1-k)^{\frac{1}{k}} \approx e^{-1}. \tag{13.48}$$

For example, with $k = 1/16$, then $(1 - 1/16)^{1/16} = 0.3461$ and $e^{-1} = 0.3679$. We can therefore approximate s_m using an exponential term as

$$s_m \approx ke^{-mk} I_{N-m}.$$ (13.49)

Therefore, the filtered image (S_N) can be approximated using a discrete convolution of temporal functions:

$$S_N = \sum_{m=0}^{N} ke^{-mk} I_{N-m} = h \otimes I.$$ (13.50)

The recursive filter operation performs low-pass filtering of the input image with a filter of the form

$$h(m) \approx ke^{-mk}.$$ (13.51)

The effect of filtering is to retain constant input signals (i.e., anatomical structures) since $h(m)$ sums to unity rather than zero and suppresses noise through image integration.

The response of a recursive filter to the time-varying bolus requires a closer look to its temporal response. An example can help with this. If m is the frame number in a sequence of images acquired with a fluoroscopic system, and if the frames are acquired at 1/30 s intervals (i.e., 30 fps), then

$$m = 30t_m,$$ (13.52)

where t_m is the time in seconds of the mth frame, and hence, the filter function can be written as

$$h(t_m) \approx ke^{-30kt_m}.$$ (13.53)

This shows that the approximated recursive filter function $h(t_m)$ has a time constant $\tau = 1/(30k)$ s (Figure 13.4). As mentioned before, Kruger found that a recursion constant having a value of $k = 1/16$ was useful for coronary arteriography. For this example, this provides a filter function with a time constant of $\tau = 16/30 = 0.53$ s. The temporal response of a recursive filter (Equation 13.53) illustrates the temporal change in filter weighting (Figure 13.5, bottom). The filtered image S_N receives the greatest weight with weight at 37% after one time constant, at 13% after two time constants, and at 5% after three time constants. When the temporal history prior to image I_N is similar to that of reflected temporal filter response, then a larger signal will be seen.

Motions that are slow relative to the filters time constant are minimally affected, while those that change too rapidly will be averaged with the earlier times. A time constant of $\tau = 0.34$ s will help to suppress motion artifacts due to breathing, will blur cardiac

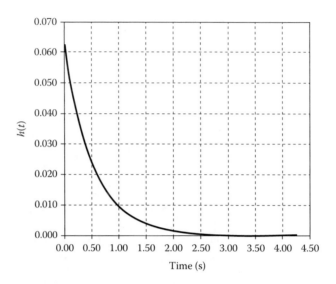

FIGURE 13.4   Recursive filter $h(t_m)$ in Equation 13.53 with $\tau = 0.53$ s.

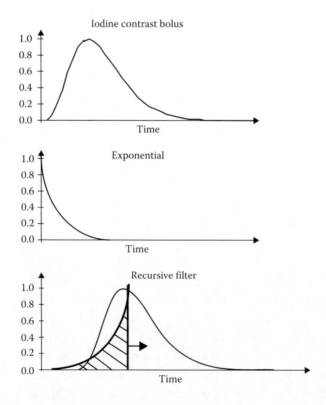

FIGURE 13.5   In recursive temporal filtering, images in a sequence are convolved with the filter and response is higher where the shape of the bolus matches the filter (bottom).

motion over a fraction of a cardiac cycle, and will average out the bolus passing through the artery for the same period of time. A shorter time constant would also be usable, but would integrate fewer frames and increase noise. If the time constant were longer, say, $\tau = 1$ s, the coronary artery would be blurred since the images would be integrated over an entire cardiac cycle.

13.8 TEMPORAL SUBTRACTION USING RECURSIVE FILTERS

As we have shown, the recursive filter retains anatomical structure. In many cases, we want to suppress anatomy as we do in mask-mode subtraction. Using two recursively filtered images with different time constants from the same data set can accomplish this. The recursive image with a long time constant can be subtracted from the recursive image with a short time constant (Figure 13.6). Objects having motions faster than that corresponding to the short time constant would be suppressed by the recursive filter, while those stationary objects retained in both recursive images would be removed by subtraction. This is the concept behind a recursive band-pass filter that simultaneously eliminates stationary anatomy, reduces noise, and suppresses the appearance of structures, such as the heart, which are moving rapidly in the image.

We begin this discussion by looking at Figure 13.7 showing the various temporal frequency components in a fluoroscopic imaging sequence following the intravenous injection of a contrast agent into the circulatory system. The delta function at zero temporal frequency represents the patient's anatomy that is stationary [obviously a gross simplification unless the person is dead or sleeping (*as during a medical imaging class*)]. The contrast bolus passing

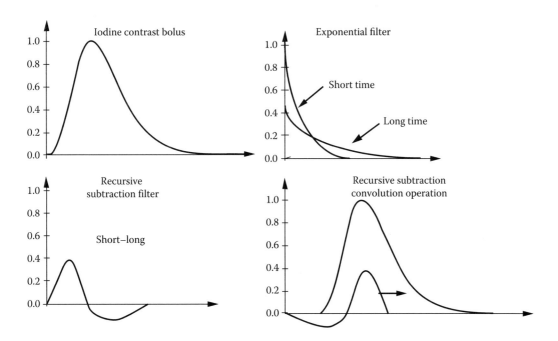

FIGURE 13.6   Subtraction following recursive filtration is equivalent to convolving (in the temporal domain) with the short-long filter.

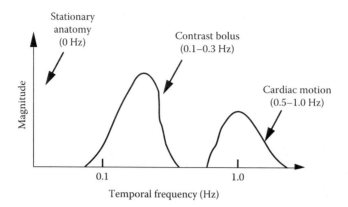

FIGURE 13.7   In recursive temporal subtraction, the filter function is formed by subtracting a filter with a long time constant with one with a short time constant. The recursive subtraction filter then is convolved with the set of images in the temporal domain. This helps to isolate the contrast bolus as it passes through the circulatory system with a given range of temporal frequencies.

through the artery is imaged over the period of 3–10 s and, therefore, is represented by the temporal frequency components in 0.1–0.3 Hz range. Finally, the subject's heart beats every 1–2 s, representing the temporal frequency in the highest range of this example.

To image the contrast bolus as it passes through the artery, but to eliminate stationary patient anatomy and suppress cardiac motion, we can design a temporal band-pass filter. This is formed using a recursive filter to obtain a low-pass filter to remove spatial frequency components above 1 Hz (where cardiac motion resides). From this, we can subtract a second low-pass recursive filter (with longer time constant) that retains only patient anatomy. The combined effect is that of a composite band-pass filter that removes patient anatomy and any structures subject to cardiac motion, but retains information about the contrast bolus passing through the artery in the intermediate frequency range. We also assume that other motions such as breathing can be controlled, for example, by breath holding.

Given the frequency distributions in Figure 13.7 for stationary anatomy, contrast bolus, and cardiac motion, it is instructive to look at the frequency response of recursive temporal filters. Using $h(t)$ derived from Equation 13.53 as the temporal domain filter response, we can calculate the magnitude $|H(f)|$ of a recursive filter as

$$|H(f)| = \left(\frac{1}{1 + \left(2\pi f \tau\right)^2} \right)^{1/2}, \tag{13.54}$$

normalized to unity output at $f = 0$. The bandwidth of this recursive filter is when $|H(f)| = \frac{1}{2}$, which is $\mathrm{BW} = \frac{\sqrt{3}}{2\pi\tau} \approx \frac{0.276}{\tau}$, so when $\tau = 0.53$ s, the BW = 0.52 cycles/s. When the spatial versions of individual recursive filters are subtracted as illustrated in Figure 13.6, the net

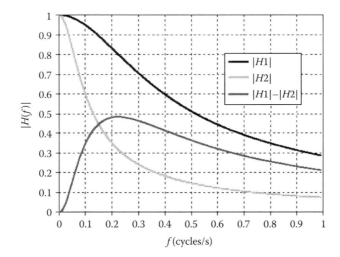

FIGURE 13.8 Magnitude frequency responses of recursive filters $H1$ and $H2$ with $\tau_1 = 0.53$ s and $\tau_2 = 2.13$ s. The $|H1| - |H2|$ band-pass filter response peaks in the 0.1–0.3 cycles/s range.

frequency response is the difference in frequency response of the filters. Figure 13.8 shows how subtraction of two recursive filters can result in a band-pass type filter. The $|H1| - |H2|$ filter has a zero magnitude at zero frequency, so it suppresses stationary anatomy. This filter has its highest magnitude for a bolus through an artery ($f = 0.1$–0.3 cycles/s) and attenuates higher frequencies such as those associated with cardiac motion.

While it is instructive to review recursive filters, medical image-processing software for temporal filtering can be achieved using high-level scientific programming applications such as MATLAB®. This topic is explored further in image-processing courses.

HOMEWORK PROBLEMS

13.1 A radiologist brings you a set of 600 digitally subtracted video frames, stored on videotape, and obtained during an angiographic study following a single injection of contrast agent. The radiologist comments that the artery is barely visible in images and asks you if you have any "image-processing tricks" to recover the images.

After viewing the tape, you realize first that the digital subtraction process has already removed the stationary anatomical information. You also discover that the images are very noisy and you decide to try a temporal matched filter to integrate the images to improve the signal-to-noise ratio. You place a large region of interest over the image and quantitate the total opacification in each frame I_i to obtain the contrast bolus curve with the individual values denoted by b_i.

(a) The matched filtering technique described in this chapter is used for processing unsubtracted images and was designed to suppress stationary anatomical information. For the image sequence given to you by the radiologist, the stationary anatomical information has already been removed by digital subtraction.

Specify a matched filter for this new image sequence (i.e., specify the coefficients h_i) that maximizes the signal-to-noise ratio of the integrated image obtained from the digitally subtracted angiograms.

(b) If $<b^2>$ is the average value of the squared opacification values,

$$< b^2 > = \frac{1}{N} \sum_{i=1}^{N} b_i^2$$

show that the signal-to-noise ratio of the integrated image obtained with the matched filter for N images frames is $\text{SNR}_{mf} = \dfrac{\sqrt{N < b^2 >}}{\sigma}$.

(c) We can compare the matched filter technique with an unweighted image integration technique where the image frames are simply added together without any constant multiplication. If the mean value of the opacification values is

$$< b > = \frac{1}{N} \sum_{i=1}^{N} b_i$$

show that this technique produces an integrated image having a signal-to-noise ratio of

$$\text{SNR}_{uw} = \frac{\sqrt{N} < b >}{s}$$

We would like to estimate the improvement in the signal-to-noise ratio produced by the matched filter in comparison to a single image and to unweighted temporal image integration. To do this, we will use the contrast bolus curve function given in the lecture notes:

$$b(t) = 0.140 t^4 e^{-0.9t},$$

which we will sample 30 times/second for 20 s, for a total of 600 frames. Furthermore, while the following calculations can be easily performed with a spreadsheet, you may also estimate the discrete samples with the continuous function $b(t)$ given earlier, and estimate the sums with integrals using the formula

$$\int X^a e^{ax} dx = e^{ax} \sum_{a=u}^{m} (-1)^a \frac{m! X^{m-n}}{(m-n)! a^{a-1}}.$$

(d) Prove that the maximum value of $b(t)$ is $b_{max} = 1$ at $t = 4.44$ s.
   Use this result to calculate the improvement in the signal-to-noise ratio obtained with the matched filter in comparison to that obtained from the best single frame in the entire image sequence.

(e) Calculate the percentage difference in signal-to-noise ratios obtained with the matched filter in comparison to unweighted temporal integration of the digitally subtracted angiograms.

13.2 We want to design an image processor capable of video-rate subtractive imaging with two recursive filters. Let I_i represent the ith (logarithmically transformed) unsubtracted image frame received by the image processor. Furthermore, the nth subtraction image S_n is defined by

$$S_n = U_n - V_n.$$

U_n and V_n are the recursively filtered images defined by

$$U_n = k_1 I_n + (1 - k_1) U_{n-1}$$

and

$$V_n = k_2 I_n + (1 - k_2) V_{n-1},$$

where k_1 and k_2 are the recursion constants, which define the temporal response of the filter.

(a) Show that the nth recursively filtered subtraction image can be written explicitly as

$$S_n = \sum_{m=0}^{n} \left[k_1 (1 - k_1)^m - k_2 (1 - k_2)^m \right] I_{n-m}.$$

(b) For large values of n (i.e., in the limit as $n \to +\infty$), assuming that the unprocessed images I_i each have a noise variance equal to σ, show that the noise variance σ_s^2 in the recursively subtracted image is given by

$$\sigma_s^2 = \sigma^2 \left[\frac{k_1}{2 - k_1} - \frac{2k_1 k_2}{k_1 + k_2 - k_1 k_2} + \frac{k_2}{2 - k_2} \right].$$

(c) Draw a block diagram for an image processor that would allow you to perform the subtractive recursive filter described in this question.

(d) The unprocessed image sequence $I(t)$ was acquired at a rate of 30 frames a second. Using the partial results given in the lecture notes, show that the recursively filtered subtraction image $S(t)$ can be represented as the temporal convolution

$$S(t) = I(t) \otimes \left[\frac{1}{30\tau_1} \exp\left(-\frac{t}{\tau_1} \right) - \frac{1}{30\tau_2} \exp\left(-\frac{t}{\tau_2} \right) \right],$$

where τ_1 and τ_2 are time constants obtained from the recursion constants

$$\tau_1 = \frac{1}{30k_1} \text{ seconds} \quad \text{and} \quad \tau_2 = \frac{1}{30k_2} \text{ seconds.}$$

(e) For $k_1 = 1/16$ and $k_2 = 1/32$, graph the temporal filter function (as a function of time) derived in part (d) of this question. Intuitively discuss the temporal response of this filter on the image sequence with respect to stationary anatomy, the contrast bolus from an intravenous injection, and cardiac motion. Calculate the noise variance of this image using the equation you derived in part (b) of this question.

V

Tomographic Imaging

CHAPTER 14

Computed Tomography

14.1 BACKGROUND

In 1917, Johann Radon showed that 2-D section images could be reformulated using mathematical transformation of projection data (i.e., using a Radon transform). Projection data are line integrals (summations of image values) recorded across an object at some angle (Figure 14.1). The connection between projection data and x-ray images was not initially obvious. However, motivation was high, since x-ray section images would have the ability to provide high contrast viewing of the body by removing interference from overlapping tissues. Later in this chapter, we will see how the x-ray projection dilemma was resolved. Even with the knowledge concerning how to make x-ray images into projections, imaging instrumentation and computing power were not able to provide this capability early on, so we had to wait many years for technology to match up with theory. By the 1960s, several research labs were able to reconstruct x-ray section images from x-ray projections acquired from physical objects, and these successes spurred intensive research on devices that could be used in medicine. In the 1970s, x-ray computed tomography (CT) was formally introduced in clinical use and was followed by rapid technological refinements. Since reconstructed images looked like the thinly sliced tissue sections used for microscopic inspection, the term "tomography," literally meaning a picture of a cut section, was adopted, and early x-ray tomographic imaging systems were called computed axial tomographic or "CAT" scanners. However, this was later dropped in favor of computed tomography or just CT. In 1979, two early researchers in the field, Allan M. Cormack and Godfrey N. Hounsfield, were jointly awarded the Nobel Prize in Physiology or Medicine for the "development of computer-assisted tomography."

Figure 14.1b is an image of the set of projections $[p_\theta(r)]$ for $\theta = 0°–180°$ for the CT image in Figure 14.1a. The top row in Figure 14.1b is at 0° and the bottom row at 180°. Columns index positions within projections (r). The sinusoidal appearance of paths formed by points 1 and 2 led us to call this image a *sinogram*. The 0° projection is a posterior view, while the 90° projection is a view from the right. Values at each $x' = r$ in the projection

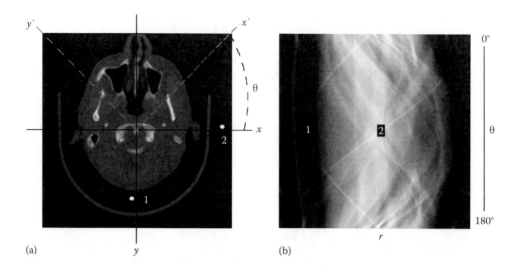

(a)   (b)

FIGURE 14.1   An x-ray CT image (a) and its projections $p_\theta(r)$ (b) presented as an image called a sinogram.

$p_\theta(r)$ are integrations across the object along a line perpendicular to the projection (i.e., line integrals along y'). The equation for mapping a point in the x-y image to a point in the r-θ sinogram (Figure 14.1) is as follows:

$$r(\theta) = r_{xy} \cdot \cos(\varphi - \theta), \tag{14.1}$$

where

$$r_{xy} = \sqrt{x^2 + y^2} \quad \phi = \tan^{-1}\left(\frac{y}{x}\right),$$

with r_{xy} as the distance from the origin to the point x, y, φ as the counterclockwise (CCW) angle from the positive x-axis to the point, and θ as the angle of the projection. The three parameters of Equation 14.1 (r, θ, and ϕ) determine the key features of the sinusoidal path followed by a point in the object:

- The amplitude of the sinusoid is determined by distance from the axis of rotation (r_{xy}).

- The phase of the sinusoid is determined from its starting phase ϕ.

- The sinusoid is theoretically fully defined over a range of θ from 0 to π or 180°.

The goal of CT imaging of the body is to obtain a set of 2-D section images, $o_n(x, y)$, from the body's 3-D object $o(x, y, z_n)$. During reconstruction, the z extent of the object (Δz) is collapsed into a 2-D section image. For simplicity, a tomographic section image will always be treated as a 2-D function, realizing that integration along the z extent of each section of the object is involved.

14.2 THEORY

Fourier transform theory provides a good theoretical approach to understand the Radon transform and more generally tomographic reconstruction of images from projections. The basis for this is provided in the following equations:

$$O(u,v) = \int\int o(x,y)e^{-2\pi i(ux+vy)}\mathrm{d}x\mathrm{d}y, \tag{14.2}$$

$$O(u,v) = \Im\{o(x,y)\} \quad (\text{shorthand notation}),$$

$$o(x,y) = \int\int O(u,v)e^{-2\pi i(ux+vy)}\mathrm{d}u\mathrm{d}v, \tag{14.3}$$

$$o(x,y) = \Im^{-1}\{O(u,v)\},$$

where the integration is performed over the domains of $o(x, y)$ and $O(u, v)$. Equations 14.2 and 14.3 (seen in earlier chapters) link these 2-D images as Fourier transform pairs. Either $O(u, v)$ or $o(x, y)$ can be calculated from the other, and therefore, each must encode a complete description of the other (Figure 14.2). Conceptually, if we are able to obtain $O(u, v)$, then we can then compute $o(x, y)$ using Equation 14.3. The following discussion focuses on how to obtain $O(u, v)$ from projections.

Figure 14.2 illustrates the correspondence between spatial and frequency domain representations of a CT section image in the head. Coordinate origins are assigned to the

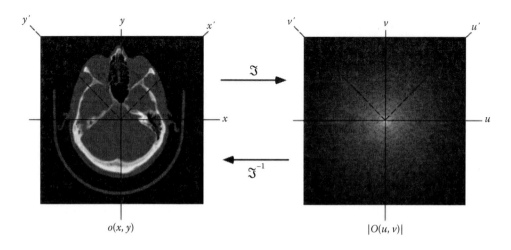

FIGURE 14.2 The spatial domain $o(x, y)$ and frequency domain $|O(u, v)|$ representations for a CT image of the head. $\mathrm{Log}_{10}|O(u, v)|$ was used in this figure to show the lower-magnitude higher-frequency terms.

center of the image arrays for both domains. This is taken to be the axis of rotation for imaging (x-ray CT and single-photon emission computed tomography [SPECT]). Image spatial coordinates (x, y) are expressed in mm and corresponding frequency coordinates (u, v) in lp/mm. While the spatial object function $o(x, y)$ is always real, the frequency object function $O(u, v)$ is usually complex. Only the magnitude of $O(u, v)$ is illustrated in Figure 14.2, but in general, $O(u, v)$ is composed of both magnitude and phase (or real and imaginary) parts.

14.2.1 Central Slice Theorem

The feature of the Fourier transform that provides great insight into computed tomography is the *central slice theorem*. This theorem states that the Fourier transform of the 1-D projection $p_\theta(r)$ in the spatial domain is identical to the profile $P_\theta(s)$ in the 2-D Fourier domain (Figure 14.3). Here, r is the distance from the origin measured in the spatial domain and s is the distance from the origin in the 2-D spatial frequency domain, with θ as the angle of the projection and profile. Calculation of the frequency domain profile from Equation 14.2 where $v = 0$ is helpful to illustrate this relationship:

$$O(u,0) = \int\int o(x,y)e^{-2\pi iux}\,dxdy,$$ (14.4)

where the order of integration can be interchanged yielding

$$O(u,0) = \int\left[\int o(x,y)dy\right]\cdot e^{-2\pi iux}dx.$$ (14.5)

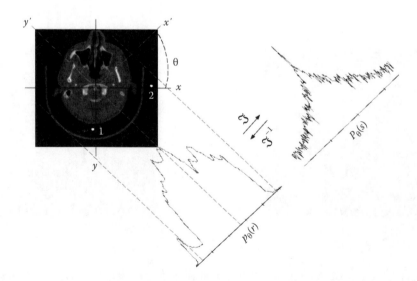

FIGURE 14.3   Formation of a projection $p_\theta(r)$ and the magnitude of its Fourier transform $P_\theta(s)$. Note that here $x' = r$.

The term within the bracket is the summation or integration of $o(x, y)$ over all y, while holding x constant (a line integral for each x). This is the projection of the object, calculated perpendicular to the x-axis. This projection can be referred to as $p_0(x)$ and this leads to

$$P_0(u,0) = O(u,0) = \int p_0(x)e^{-2\pi iux}\mathrm{d}x. \tag{14.6}$$

Equation 14.6 shows that the profile $P_0(u, 0)$ in the 2-D frequency domain at $v = 0$ corresponds to the 1-D Fourier transform of the projection $p_0(x)$ in the spatial domain. Both the projection and the profile correspond to data acquired at $\theta = 0$. A more general equation for the frequency domain profile is

$$P_\theta(s) = \int p_\theta(r)e^{-2\pi irs}\mathrm{d}r. \tag{14.7}$$

The profile $P_\theta(s)$ is a central profile, because its origin coincides with the origin in the frequency domain. Equation 14.7 can be shown to be true at any angle. As stated previously, if we acquire sufficient data to determine the 2-D Fourier transform of the object $O(u, v)$, then we can reconstruct it using Equation 14.2. *The objective is to fill in the 2-D Fourier space by acquiring a sufficient number of projections about the object, and this can be done using projections spanning 0°–180°.* The following points summarize the theoretical requirements and basis for tomographic reconstruction in computed tomography:

- Projections $p_\theta(r)$ are summations along a line, or line integrals of the object function values, at projection angles θ.

- For each spatial domain projection $p_\theta(r)$, there is a corresponding frequency domain central profile $P_\theta(s)$.

- If a sufficient number of spatial domain projections are acquired, then a sufficient number of central profiles $P_\theta(s)$ can be calculated to fill the 2-D Fourier domain. An inverse Fourier transform can be used to calculate $o(x, y)$ using Equation 14.3.

These points help to explain how a tomographic image can be calculated from projections. There are various methods to acquire projection data for x-ray CT and nuclear medicine PET and SPECT images, but all are based on this mathematical description of computed tomography. Several reconstruction methods will be discussed later in this chapter.

14.2.2 Definition of Projection

A challenge for CT image acquisition is to acquire projection data from a section of the object that effectively fills the 2-D frequency domain. A strict requirement is that the acquired data conform to the definition of a projection. A projection can be mathematically defined using a delta function as

$$p_\theta(r) = \int\int o(x,y)\delta(r - x\cos(\theta) - y\sin(\theta)\mathrm{d}x\mathrm{d}y. \tag{14.8}$$

To help understand the role of the arguments in the delta function, we need the correspondence between locations in natural (x, y) and rotated (x', y') coordinate systems. The mathematical relationship for a CCW rotation of angle θ about the origin $(x, y = 0, 0)$ is determined using the following transform matrix:

$$\begin{bmatrix} x' \\ y' \end{bmatrix} = \begin{bmatrix} \cos(\theta) & \sin(\theta) \\ \sin(-\theta) & \cos(\theta) \end{bmatrix} \begin{bmatrix} x \\ y \end{bmatrix}. \tag{14.9}$$

From Equation 14.9, we see that the argument of the delta function in Equation 14.8 involves $x' = x\cos(\theta) + y\sin(\theta)$. Therefore, the integration in Equation 14.8 is constrained to be along a line $r = x'$ by $\delta(r - x')$, which is parallel to the y' axis (see Figure 14.3). The coordinate origin for x-ray CT is the axis of rotation or the axis about which projections are acquired. Equation 14.9 can be used for backprojection and reprojection, since given x', y', and θ, we can calculate x and y. During projection for each x' (analogous to r), we sum along y' to calculate $p_\theta(r)$. For backprojection, we divide the projection data $p_\theta(r)$ equally along y'.

In x-ray and nuclear medicine tomographic imaging (SPECT) projections, $p_\theta(r)$ are usually acquired by rotating the imaging device through a series of angles (θ) about the object. Since the projection at angle θ should be identical to that at angle $\theta + 180°$, only $180°$ scanning is required. This is confirmed by the fact that the 2-D Fourier space is completely filled with profiles spanning $180°$. For various reasons, the scan angle extent is usually larger than $180°$. However, angular extent smaller than $180°$ will not completely fill the 2-D Fourier space of the object and leads to reconstruction errors. *Note*: Undersampling of the 2-D Fourier space can be partly compensated by interpolating values between missing profiles or reducing the highest frequency used during reconstruction, but both lead to reduced spatial resolution. Additionally, for SPECT, opposing projections are not equal due to differences in attenuation, so opposing projections, each spanning $180°$, are used to estimate corrected projections.

14.2.3 Line Integrals

As stated in the introduction, a projection must be composed of line integrals (i.e., summation) of object values. This requirement is especially important in x-ray CT, since raw projection data are the intensity of the x-rays transmitted through the object, not an integration of object values. To understand how x-ray CT projection data are converted to a proper set of line integrals, the characteristics of the raw x-ray projection data, need to be analyzed. The x-ray intensity at a location r in a projection is modeled as follows:

$$I_\theta(r) = I_0(r)e^{\int\int -\mu(x,y)\delta[r-x\cos(\theta)-y\sin(\theta)]dxdy}, \tag{14.10}$$

where

$I_0(r)$ is the intensity at r without the object

$\mu(x, y)$ is the linear attenuation coefficient at object location x, y

Dividing both sides of Equation 14.10 by $I_0(r)$ and taking the natural logarithm leads to an equation for the integral of linear attenuation coefficients, that is, projection $p_\theta(r)$ along a line defined by r:

$$p_\theta(r) = \ln\left(\frac{I_0(r)}{I_\theta(r)}\right) = \int\int \mu(x,y)\delta(r - x\cos(\theta) - y\sin(\theta)dxdy. \tag{14.11}$$

For x-ray CT, raw projection data are converted logarithmically into proper projection data using Equation 14.11. *Inspection of this equation shows that in x-ray CT the values computed during reconstruction are linear attenuation coefficients* $\mu(x, y)$. Unlike x-ray CT, for SPECT the raw projections do not require such conversion and the computed values in SPECT images are activity concentration/voxel (Bq/cc).

14.2.4 X-Ray CT Number Standardization

While x-ray CT directly calculates images with pixel values that are linear attenuation coefficients, a different scheme was devised to help standardize CT numbers. Hounsfield suggested that it would be useful to report CT values as relative attenuation with the attenuation coefficient of water being the reference value. This led to the following equation for CT numbers:

$$CT\# = \frac{\mu - \mu_{water}}{\mu_{water}} \times 1000. \tag{14.12}$$

Inspection of these CT#'s shows that the $CT\#_{water} = 0$ and $CT\#_{air} = -1000$. Since the linear attenuation coefficient of fat is less than that of water, the CT number of fat is negative. Most other soft tissues are positive, while that of dense bone can be as high as 3000. The adoption of CT numbers has helped to standardize CT image values across machines and hospitals. Importantly, x-ray CT numbers can be less sensitive to changes in kVp and beam filtration than are linear attenuation coefficients. Can you explain why?

14.2.5 Reconstruction by Backprojection

While there are several approaches to determine $o(x, y)$ from $O(u, v)$, the most basic, though not mathematically correct, is called simple backprojection. Simple backprojection is an attempt to redistribute line integral data into the object. Backprojection uniformly redistributes the line integral values within a circle of diameter equal to the length of the

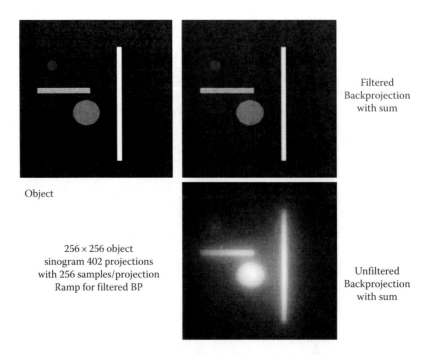

Filtered
Backprojection
with sum

Object

256 × 256 object
sinogram 402 projections
with 256 samples/projection
Ramp for filtered BP

Unfiltered
Backprojection
with sum

FIGURE 14.4 Filtered backprojection image is a faithful reproduction, while simple (unfiltered) backprojection image is blurred.

projections. This circle is called the scan circle. *The entire object must reside in the scan circle to be correctly reconstructed; otherwise, it will not be sampled at each projection.* Simple backprojection (unfiltered backprojection in Figure 14.4) fails to correctly reconstruct the object from its projections. This is easy to understand from the following example:

If a point-like object (centered in the scan circle) is imaged, each projection will be identical, having data only at its center. Backprojection of the first projection divides its value (sum of image date) equally along a line passing through the origin. A similar uniformly filled line will be backprojected for each projection angle. During each backprojection, new values are added to previous values. At the center of the image, each backprojection contributes to the sum; however, backprojections also sum further away from the center. Therefore, the image of the point source reconstructed using simple backprojection is spread out. This image is in fact the point spread function (PSF) of image formation using simple backprojection (Figure 14.5). This PSF diminishes with distance from the center following a $1/r$ trend, and results from the overlapping of backprojected lines, where all lines overlap at the origin.

There is another way to look at the problem associated with simple backprojection. If simple backprojection worked properly, then reprojection from the backprojected image should produce projections identical to those from the real object. This cannot happen due to the overlapping of different projections, which produce nonzero values for line integrals away from the center during reprojection. The only solution to this dilemma is for the

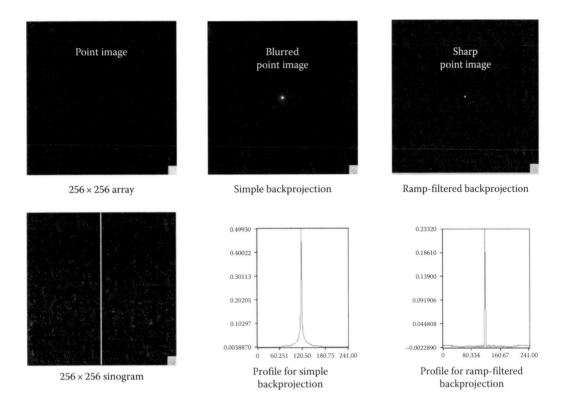

FIGURE 14.5   Comparison of PSFs for simple and ramp-filtered backprojection methods.

projection data to be modified to include both positive and negative values to remove such. This will become obvious soon.

While simple backprojection forms a blurred version of a point object (Figure 14.5, middle), a special ramp-filtered backprojection approach (Figure 14.5, right) produces a nearly perfect replica of the point object. The source of this problem can be seen by inspection of the PSF of simple backprojection (Figure 14.5). The profile through this spread function reveals that the simple backprojection psf(r) drops off as $1/r$. There are two approaches to remove the blurring introduced by this $1/r$ broadening of psf(k). One method to correct for the blurring is to deconvolve the backprojected image with an appropriate function in the spatial domain. However, this is not the most straightforward approach, and a Fourier domain approach is usually preferred. Blurring in the spatial domain is described mathematically as follows:

$$i(x, y) = o(x, y) \otimes\otimes \text{psf}(x, y), \quad (14.13)$$

where psf(x, y) = k/r and $r = (x^2 + y^2)^{1/2}$. The corresponding equation in the Fourier domain is

$$I(u, v) = O(u, v)\, \text{STF}(u, v), \quad (14.14)$$

where $STF(u, v) = K/\rho$, $\rho = (u^2 + v^2)^{1/2}$. This STF shows a net reduction in frequency response following an inverse ρ trend for simple backprojection. For realistic PSFs, neither k/r nor K/ρ approach ∞ as their denominators approach zero. In fact for real CT, the value of K/ρ is equal to an integral about k/r near $r = 0$ (determined by voxel size). We will see how this can be calculated for discrete CT data later. The drop-off in frequency response in $I(u, v)$ can be correctly compensated by multiplication of $I(u, v)$ by the inverse of STF, which is ρ/K, to calculate $O(u, v)$ as follows:

$$O(u, v) = I(u, v) \cdot \rho/K. \tag{14.15}$$

The ρ/K term is a straight line with zero intercept as a function of ρ, and due to this appearance is referred to as a "ramp" filter. Similarly, the processing indicated in Equation 14.15 is called "ramp filtering." The ramp filter compensates for the loss in high-frequency response due to the oversampling of low frequencies and likewise undersampling of high frequencies associated with summation and backprojection.

Ramp filtering is generally done on a projection-by-projection basis to provide filtered projections as follows:

$$p'_\theta(r) = \mathfrak{I}^{-1}[P_\theta \cdot (s) \cdot |s|/K, \tag{14.16}$$

where $|s|$ is the frequency equivalent of ρ from Equation 14.15. Backprojection after filtering with a ramp filter is called "filtered backprojection." Its effect is to create an almost perfect reconstruction (Figure 14.5).

An intuitive description of the requirement of a ramp filter for proper CT reconstruction from projections follows from inspection of $p_\theta(r)$, $P_\theta(s)$, and $PSF(u, v)$ for a point object $\delta(x, y)$. For a perfect CT imager, that is, no unsharpness, each $p_\theta(r)$ will be a 1-D delta function $\delta(r)$ and its Fourier transform, $P_\theta(s)$, will be a constant $= 1$. To model simple backprojection in the u, v domain, each $P_\theta(s)$ must be added to the u, v domain at the proper angle. This means that the response at the origin [$PSF(u = 0, v = 0)$] will be equal to N_p where N_p is the number of projections overlapping at the origin, since the central slice theorem states that all $P_\theta(s)$ pass through the origin. Also, $PSF(u, v)$ will fall off as $1/|s|$ moves away from the origin. This is because at each radial distance from the origin, $2N_p$ projections are summed along a circle of circumference $= 2\pi\rho$, so the magnitude decreases with increasing radius as $N_p/(\pi\rho)$ ($|s|$ and ρ are used synonymously). To achieve a constant frequency response (the correct Fourier transform of $\delta(x, y)$), $PSF(\rho)$ must be multiplied by a frequency compensation term like $(\pi\rho)/N_p$. While this description does not exactly match that in Equation 14.15, the difference decreases with normalization of the final data.

The 2-D Fourier transform of $\delta(x, y)$ should be uniform across the u, v frequency space. The need to increase magnitude as a function of frequency stems from the fact that while each profile can have a uniform frequency response, when summed into the

2-D Fourier space, the net magnitude decreases following $1/\rho$. Compensation for this high-frequency drop-off results in "equalized" frequency response and explains the necessity for the ramp filter. Since it is simpler to perform the filtering in 1-D, Equation 14.16 is usually used. Also, when using this technique, reconstruction can be done during data acquisition, where each acquired projection is filtered and backprojected as soon as it is acquired. Reconstructed slice images are available immediately following the acquisition of a complete set of projection data. Filtered backprojection implements the Radon transform.

It is certainly possible to reconstruct $o(x, y)$ directly from $O(u, v)$ by taking the inverse Fourier transform bypassing the need for backprojection. However, unlike filtered backprojection where backprojection can be done immediately following acquisition of each new projection, the 2-D inverse Fourier transform method can be applied only after all projections have been transformed to Fourier space, that is, the end of the scan. Early systems were very slow and filtered backprojection soon became the preferred approach. It should be noted that filling in $O(u, v)$ from Fourier transform of projections must be done carefully to avoid summing of overlapping projections. One way to do this is to sum and fill normally, but save a buffer of how many entries were made, and correct to the average values once all projections are acquired.

Projection data do not have to be acquired using parallel lines for integration. For improved geometrical efficiency, fan beam geometry is commonly used in x-ray CT. The fan beam line integrals can be sorted into parallel projections before processing or processed using a different filter. However, all discussions in this chapter focus on parallel beam-type projections and reconstructions.

14.3 PRACTICAL CONSIDERATIONS

14.3.1 Number of Samples per Projection

The value of each sample point in a projection is a line integral across the object. Each sample is acquired by a single or grouping of detectors. The number of samples per projection, N_s, can be estimated using Shannon's sampling theorem. Recall that this theorem specifies that the sampling frequency $f_s \geq 2f_{max}$ where f_{max} is the highest frequency or limiting bandwidth of the imaging system. This sampling frequency is necessary to avoid aliasing. The number of samples per projection based on the sampling theorem is

$$N_s = 2\ [\text{samples/lp}] \times f_{max}\ [\text{lp/mm}] \times \text{FOV}\ [\text{mm}], \tag{14.17}$$

$$N_s = 2f_{max}\text{FOV}\ [\text{samples}].$$

Example 14.1

N_s for X-Ray CT
For CT imaging $f_{max} \sim 1$ lp/mm and FOV = 256 mm, so $N_s = 512$ samples.

Example 14.2

N_s for SPECT

For SPECT, f_{max} ~ 1 lp/cm, and N_s = 50 for a 25 cm FOV (would likely use N_s = 64). This would increase to N_s = 80 for a 40 cm FOV (would likely use N_s = 128).

14.3.2 Number of Projections

The number of projections (N_p) is calculated to ensure equivalent sampling around the scan circle in u, v space. This is accomplished if sampling along the circumference of the scan circle (one sample per projection) in u, v space matches sampling along the diameter. The number of samples in a projection N_s is the same as the number of samples along the diameter of the scan circle in Fourier space (ranging from $-f_{max}$ to $+f_{max}$). To maintain this sample spacing, we need to have πN_s samples around the scan circle's circumference. Since each projection's Fourier profile provides two points in u, v space along this circumference, the number of projections needed is one-half the πN_s value or

$$N_p = \pi/2 \cdot N_s \text{ [projections]}. \tag{14.18}$$

Figure 14.6 illustrates the improvement in a reconstructed point object following the guideline given in Equation 14.18. N_p is just a multiple of N_s, so it can be calculated similarly from f_{max} and the FOV:

$$N_p = \pi f_{max} \text{FOV}. \tag{14.19}$$

The total number of line integrals acquired is equal to the number of samples per projection times the number of projections = $N_s N_p$. The following table summarizes this for several imaging system configurations:

Application	N_s	N_p	# Line Integrals
Nuc Med	64	101	6,464
Nuc Med	128	201	25,728
X-ray CT	512	804	411,648
X-ray CT	1024	1608	1,646,592

14.3.3 CT Filter Response

Though the mathematical form of the "ramp" filter for theoretically correct CT reconstruction is well known, in most cases, the filter must be modified to reduce output at higher frequencies where the SNR can be very low. This is accomplished using a low-pass filter. Though the bandwidth of the low-pass filter is user selectable most x-ray CT consoles only provide options such as high-, medium-, and low-resolution filters. PET and SPECT systems

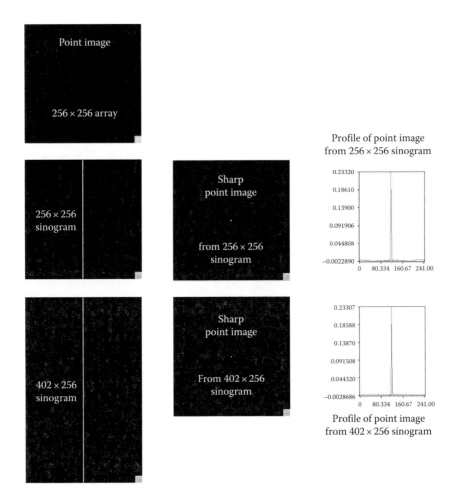

FIGURE 14.6 Bottom shows improvement in background around reconstructed point when the number of projections $N_p = \pi/2 \cdot N_s$ where N_s is the number of samples per projection.

generally provide a wider range of filters. A common low-pass filter is a Butterworth filter (Figure 14.7). It is a two-parameter filter with frequency response as follows:

$$\text{Butterworth}\,(u) = \frac{1}{1 + (u/u_0)^{2N}}, \tag{14.20}$$

where
 u_0 is the frequency where filter output = 1/2 (called the *bandwidth* of the filter)
 N is the order of the filter

The filter magnitude = 1 for $u = 0$ and approaches zero large values of u. The steepness of the filter response around u_0 increases as N is increased. The selection of N is based on frequency response needs above and below the designated bandwidth.

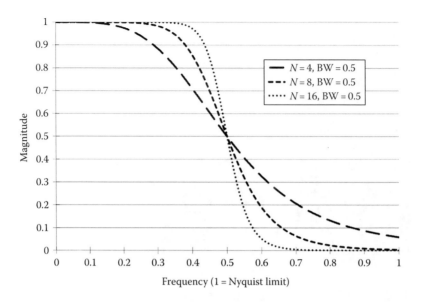

FIGURE 14.7   Frequency response for several Butterworth type filters used with SPECT and PET.

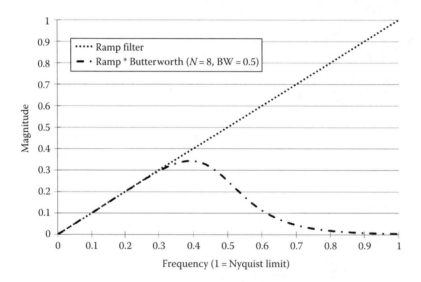

FIGURE 14.8   Net CT filter response is ramp (dotted) multiplied by a Butterworth filter.

The net CT image frequency response is determined as the product of the ramp filter and the user-selected low-pass filter (Figure 14.8). The net filter response peaks somewhere below u_0, tracks the ramp at lower frequencies, and tends to approach zero at higher frequencies.

Figure 14.9 shows the MTF and noise of a tomographic imager prior to application of the net CT filter. It demonstrates the need to taper the output of the filter at high frequencies where SNR is poor, while following the theoretical ramp at lower frequencies.

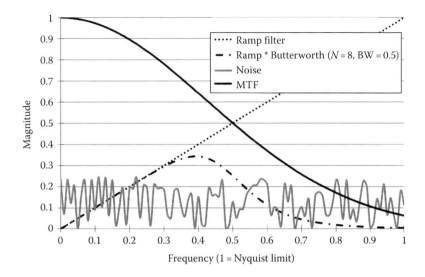

FIGURE 14.9   Noise and MTF indicate the bandwidth to use for filtered ramp.

Low-pass filtering is critical in nuclear medicine (SPECT and PET), because the projection data are often very noisy due to the limited number of quanta per voxel. To deal with this, low-pass filter bandwidth is set at approximately 1/2 the Nyquist limiting frequency, as illustrated in Figure 14.9. While such low-pass filters improve the SNR in reconstructed images they degrade resolution, with response above u_0 more attenuated.

In x-ray CT, similar low-pass filtering is available, but u_0 appears to be well above 1/2 of the Nyquist limit, since image detail is maintained with good SNR over a broader range of frequencies for most studies. Smoother images result for u_0 near 1/2 Nyquist limit with sharpness improving as u_0 is increased. The higher SNR at any given frequency for x-ray CT versus SPECT and PET systems is due to the much larger number of x-ray quanta per voxel acquired in x-ray CT.

14.3.4 Beam Hardening

A problem with using the naturally polyenergetic x-ray beam for CT is that the beam will be harder (mean energy higher) for thicker body regions, due to the longer attenuation path. A common example of this is when imaging a cylindrical phantom (Figure 14.10). Note that higher mean energy means less attenuation and indicates a smaller average linear attenuation coefficient (μ) for x-rays passing through center of the phantom. The profile of a reconstructed CT image will be lower in the middle, since μ is proportional to CT#. The cupping seen in the graph of Figure 14.10 is the result of beam hardening in x-ray CT images.

There are several ways to resolve this beam-hardening problem. A common method is to preharden the x-ray beam with a thicker in-beam aluminum filter so that the additional hardening caused by the patient leads to a smaller percentage change in the mean x-ray beam energy. This approach is supplemented by the use of a higher kVp beam

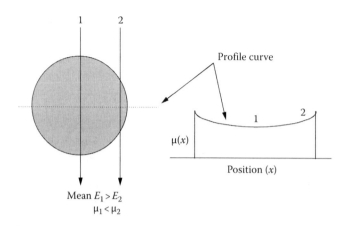

FIGURE 14.10 X-ray CT beam hardening for a cylindrical phantom of uniform attenuation illustrating dip in attenuation coefficients calculated near the center of the projection profile curve.

(125–130 kVp) than for routine projection radiographic imaging. A second method to reduce the beam-hardening effect seen in Figure 14.10 is to use a "bow-tie"-shaped aluminum filter to preharden the periphery (i.e., ray 2) more than the center of the beam (ray 1). This can be effective for cylindrical objects, but is not acceptable as a general solution. A third method, and one not used much anymore, is to surround the object with a water bag. This was mostly used to reduce the dynamic range of x-ray intensity between detectors in the middle of the FOV and those at the periphery. Early CT detectors were limited in dynamic range, the ratio of highest to lowest x-ray intensity seen by the detectors.

A more analytical approach to correct for periphery-to-center beam hardening is based on a measured ratio of I_0 (no-attenuation signal) to I (attenuated signal). The logarithm of the ratio of I_0 to I, if mean energy does not change (i.e., no beam hardening), should be a line with slope = μ and intercept = 0 when plotted against diameter of a cylindrical water phantom (Figure 14.11). This follows from the simple equation describing attenuation of x-rays:

$$\ln(I_0/I) = \mu d. \tag{14.21}$$

Data acquired with varying diameter d are used to correct measured values of $\ln(I_0/I)$ to theoretical values using a look-up table. In this example, a value of $\ln(I_0/I) = 4.8$ would be mapped to a value of 5.6. This is a reasonably good method to correct the periphery-to-center beam-hardening problem for soft tissues and near cylindrical body sections. Some form of this correction is used on all x-ray CT machines.

Another beam-hardening problem often seen in head CT images comes from large differences in attenuation for rays traversing bone versus soft tissue. This beam hardening cannot be corrected completely without some form of iterative processing. The result is streak artifacts near sharp edges of bone.

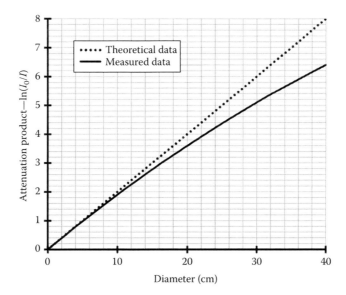

FIGURE 14.11 X-ray beam-hardening correction scheme adjusts measured attenuation products to match the theoretical values.

14.4 OTHER CONSIDERATIONS

Attenuation correction for SPECT and PET is required to provide true line integrals for projections. PET attenuation correction is quite good though SPECT attenuation correction is sometimes poor. Both can use 360° data acquisition, and this is needed for good attenuation correction. PET acquires multiple slices simultaneously without rotation of the detectors. SPECT acquires multiple slices in one rotation of detectors. X-ray CT now can acquire multiple sections with high-speed helical scanners.

HOMEWORK PROBLEMS

P14.1 Calculate the projections ($P_\theta(r)$ from Equation 14.11) at 0° and 45° for a square object 10 cm on each side, if $\mu = 0.2$ cm^{-1}. Use the coordinate system given in Figure 14.2.

P14.2 You are working on a research project to build a micro-CT system with x, y, and z spatial resolution of 10 microns and a field of view of 5 cm. The system uses a pencil beam that is scanned across the object and then the object is rotated. The object is translated (stepped) in the z direction and the process is repeated. What values would you select for each of the following and show how you arrived at your conclusion:

(a) Dimensions of x-ray beam

(b) Distance between steps in the z direction

(c) Number of samples per projection

(d) Number of projections

(e) Image matrix size

(f) Form of the reconstruction filter

P14.3 A simple formula can be used to estimate the form of the ramp filter for filtering the Fourier transform of projections. Given that $\text{Ramp}(u) = 1/N_p$ at $u = 0$ and $\text{Ramp}(u) = 1$ at $u = u_{max}$, show that

$$\text{Ramp}(u) = \left[\frac{1 - \dfrac{1}{N_P}}{\dfrac{N_s}{2}}\right]|u| + \frac{1}{N_P}.$$

P14.4 The mean linear attenuation coefficient for a 40 cm water phantom is 0.2 cm^{-1}. What is the dynamic range of x-ray intensity seen by the CT detectors? Can this range be properly recorded with a 16-bit binary data format?

P14.5 You have been asked to determine the center of rotation (COR) of a SPECT camera. Explain how you might do this using modified Equation 14.1. Also, suppose that the SPECT camera is wobbling about its COR during rotation (maybe because of gravity), how can you detect this problem?

CHAPTER 15

Single-Photon Emission Computed Tomography

15.1 BACKGROUND

Single-photon emission computed tomography (SPECT) systems are available with single and multiple camera heads. Dual head systems are common, where camera heads are set 180° apart on a rotating gantry, a configuration that is naturally counterbalanced. Spatial resolution decreases with distance from the face of the camera's collimators due to the overlapping of the field of view (FOV) of adjacent collimator channels. Most collimators for SPECT are parallel-channel designs with longer channels than used for planar imaging, since resolution does not diminish as rapidly with distance for longer channels. Some SPECT systems use body contouring, where the camera follows an outline of the body as it rotates about the patient, in order to keep the collimator close to the body. Iterative reconstruction techniques can incorporate models of collimator resolution changes with distance such that modern SPECT systems can provide spatial resolution of ~10 mm at depth. Since sources deep within the body are attenuated more than superficial ones, attenuation correction must be included in the reconstruction process. Attenuation correction methods are based on estimating attenuation for each projection. For combined SPECT/CT systems, the attenuation coefficients are estimated from the CT image with attenuation coefficients adjusted to the energy of the radionuclide used (mostly 140 keV for Tc-99(m)).

Spatial resolution, contrast, uniformity, and sensitivity should be routinely evaluated for nuclear medicine imaging systems, with more strict criteria for SPECT systems than planar gamma cameras. However, many important assessments can be performed using only planar images.

15.2 SPATIAL RESOLUTION

15.2.1 Intrinsic Resolution

It is important to assess the intrinsic spatial resolution (without collimator) of each camera head. Routine testing (quality control) of spatial resolution often uses a 4-quadrant bar phantom to assess limiting spatial resolution (Figure 15.1). Here, the bar width and spacing is identical within each pattern and the smaller bars represent higher spatial frequencies. For large field of view (LFOV) cameras, the bar phantom image is acquired with a minimum of two million counts. The image should be acquired such that these counts are within the bar pattern; however, in Figure 15.1a, a large proportion of the counts were outside the range of the bar patterns (an example only). Gamma camera vendors usually supply a lead device to shield out these peripheral counts. In this example, all bars were visualized, so the limiting intrinsic resolution spatial resolution could not be determined visually, since it would have occurred with bars smaller than those of the phantom. For visual only assessment of limiting spatial resolution, the test should be done with bar phantoms having small bars that cannot be visualized and preferably more than four bar patterns.

While visual assessment is common for gamma camera quality control, a more quantitative approach can be used to track changes in spatial resolution and determine when to take corrective action. Profile graphs spanning multiple bars in the 256 × 256 pixel image show different periods/frequencies and amplitudes for two bar patterns (Figure 15.2, bottom). Note the variability in each pattern's amplitude in the profile graphs.

A quality control bar phantom image such as that in Figures 15.1 and 15.2 can be used to assess spatial resolution reliably and quantitatively. Quantitation is based on

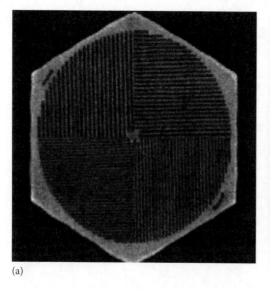 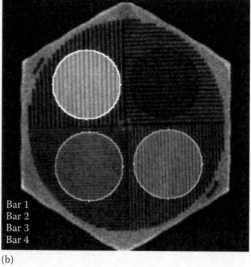

Bar 1
Bar 2
Bar 3
Bar 4

(a)

(b)

FIGURE 15.1 **(See color insert.)** Bar phantom image for LFOV camera head (a) and ROI for assessing change of contrast with decreasing bar size (b).

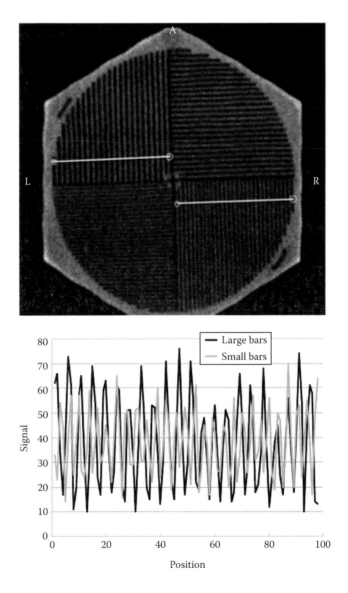

FIGURE 15.2   Profile graphs for larger and smaller bars.

statistical measures (mean and variance) within regions of interest (ROIs) for each bar pattern (Figure 15.1b). The ROIs must be properly sized and positioned to stay within the bars. Circular ROIs are preferred, since these cover approximately the same bar area regardless of phantom orientation. The variance measured within the ROIs is due to three sources: (1) the amplitude of the bar pattern, (2) quantum statistical variance (equal to the mean value), and (3) variance due to spatial nonuniformity of the gamma camera (Figure 15.4). *The variance due to the bar patterns is estimated by subtracting quantum and camera variance from the measured variance.*

The contrast (C) for periodic signals such as those in the profile curves in Figure 15.2 is the amplitude divided by the mean value, but the amplitudes vary significantly across the profiles, making it difficult to estimate amplitude accurately. *However, the standard deviation (SD) of the signal in the profile curve is easy to determine and is related to the signal amplitude.* Additionally, the SD and mean values for profiles can be calculated without explicitly determining amplitude, using all of the profile's data, not just peak and valley values. Importantly, the relationship between mean values and SDs for profiles also holds for the ROIs used to assess these statistics within the different bar patterns. Therefore, for the ROIs, the ratio (estimated SD)/(mean value) can be used as a SD-based measure of contrast (C_{SD}), where the SD has been corrected for random and spatial uniformity sources of variance. The C_{SD} values calculated for the bar phantom in Figure 15.1b are 0.44, 0.38, 0.32, and 0.21 for the largest to the smallest bars. The C_{SD} values (0.46 and 0.32) calculated for the profiles in Figure 15.2 were similar to those using only the ROIs. While results are similar, reproducibility is poorer when using the profiles due to the smaller number of samples values (quantum variance), variations due to positioning (location and orientation) of profile lines, and camera nonuniformity that is not usually assessed for a profile line. Therefore, the large circular ROIs are recommended for this assessment.

C_{SD} values by bar size can be recorded and tracked weekly to monitor for changes. Loss of spatial resolution will affect smaller bars (higher-frequency sinusoidal pattern) more than larger bars, adding to the specificity of this approach. A reduction in C_{SD} of more than 10% for the smaller bars (the ones with contrast = 0.21) could be designated as a threshold for action. However, actual values and thresholds for action need to be evaluated for each gamma camera. The utility of this approach is that the ROIs can be easily positioned within each bar pattern as compared with other methods that require careful alignment for determining profile values. Additionally, many different points in the bar phantom image are analyzed, reducing confounds associated with pixel size and statistical noise.

Since it is important to assess spatial resolution for both horizontal and vertical lines, and to do this in each quadrant, this can be done by periodically rotating the quadrant phantom to the four 90° orientations, then flipping to swap, and repeating to swap horizontally and vertically oriented patterns. This need not be done routinely, but can be used to provide baseline measures of the four patterns in each quadrant of FOV for both horizontally and vertically oriented bars.

Intrinsic gamma camera spatial resolution can also be evaluated from an image of an edge of a lead plate formed using a small point source at a distance of ~5 × FOV of the gamma camera. Mean counts/pixel due to background radiation (assessed behind the lead plate) must be subtracted from each pixel. A profile curve across the edge provides an edge-spread function (ESF). Multiple profiles can be acquired, aligned, and averaged to improve statistics. Setting the maximum value to unity normalizes the ESF. The derivative of the ESF is the LSF. To manage random noise, the LSF is usually fit with a Gaussian function and where $FWHM_{raw} = 2.35\sigma$ with σ determined from the fitted LSF.

Unlike the approach with a bar phantom using circular ROIs, the pixel size for edge imaging must be small compared to the gamma camera's FWHM. A pixel spacing of 1/5th–1/10th the FWHM is adequate, but that may not be possible. For the gamma camera used for Figure 15.2, this would require approximately 2×–3× as many pixels in each direction, or 4×–9× as many total pixels, and most users would not routinely do this, *but physicists and engineers when making this assessment should use the highest detail matrix.* The effect of pixel size can be partially corrected using

$$\mathrm{FWHM_{net}} = \left(\mathrm{FWHM_{raw}^2} - \Delta^2\right)^{1/2}, \tag{15.1}$$

where Δ is the pixel width.

15.2.2 Extrinsic Resolution

The extrinsic spatial resolution of a gamma camera is measured with the collimator on (**EXON** to help remember). This in an overall measure of system spatial resolution and should be evaluated for each collimator at two distances, 0 and 10 cm (or 20 cm for tomographic systems) from the collimator face. The bar phantom used for intrinsic resolution testing is not appropriate for use with the collimator on (due to interference between the periodic collimator channel spacing and the bars at 0 cm distance), so visual assessment for QC is not reliable at that distance. Extrinsic gamma camera spatial resolution can be evaluated using a radioactive edge, line, or point sources, but line and point sources are more common. A microhematocrit capillary tube with inner diameter <1 mm containing Tc-99(m) can be used for acquiring a line-source image. A Co-57 spot marker (1″ diameter acrylic with 3.0–3.5 mm diameter sealed source) can be used to acquire a point source image. A line spread function (LSF) or point spread function (PSF) is calculated from these images and the FWHM and FWTM determined as an indices of system spatial resolution. The larger diameter of the Co-57 marker will lead to an overestimation of the FWHM (especially at 0 cm), but this can be partially corrected for using

$$\mathrm{FWHM_{net}^2} = \mathrm{FWHM_{raw}^2} - d^2, \tag{15.2}$$

where d is the diameter of the Co-57 marker source.

15.2.3 Collimator Resolution

The spatial resolution due to collimator alone can be estimated using the following

$$\mathrm{FWHM_{extrinsic}} = \left(\mathrm{FWHM_{collimator}^2} + \mathrm{FWHM_{intrinsic}^2}\right)^{1/2} \tag{15.3}$$

and solving for $\mathrm{FWHM_{collimator}}$. This and the equation for correcting for marker (15.2) size are based on the assumption that the spread functions are uncorrelated and approximately Gaussian.

15.2.4 Resolution by Direction and Position

For planar imaging systems, spatial resolution is generally calculated for both vertical (y) and horizontal (x) directions. These may differ across the FOV, and detailed studies can be performed to test such. For example, a bar phantom can be positioned such that both horizontal and vertical bars of each bar pattern are possible for each quadrant of the gamma camera. Then, C_{SD} could then be compared by quadrant. A test such as this would not need to be performed weekly but may be indicated annually, or after major upgrades or repairs.

Since SPECT produces 3-D tomographic images, there are more directions of interest in spatial resolution (including between slices). Additionally, since collimator resolution varies with distance, it is important to assess these at the center and periphery of the reconstructed section images. Point or line sources can be used to assess spatial resolution for SPECT. The acquisition parameters (number of samples per projection, number of projections, and total counts) should be held constant for repeat testing of spatial resolution. Also, tomographic spatial resolution should be tested using only the "ramp" reconstruction filter. This avoids the effect of low-pass filters on the high-frequency response. Finally, sufficient counts should be obtained to minimize effects of quantum noise. It may also be beneficial to measure reconstructed spatial resolution with low-pass filters commonly used in clinical studies. Further information regarding methods for assessing spatial resolution can be found in AAPM Report 52 *Quantitation of SPECT Performance*.

15.3 CONTRAST

Subject contrast in nuclear medicine images was a topic in Chapter 4. Scatter within the subject reduces image contrast. Since scattering reduces the energy of gammas, the effect of scatter can be moderated by only accepting counts for events when the energy is within ±10% of the desired gamma energy (±14 keV for 140 keV photons of Tc-99(m)). This energy range preferentially accepts primary radiation while rejecting larger-angle scatter. Contrast at different distances can differ with different collimators, so it should be tested with consistent energy window settings and collimators. A phantom with two chambers containing known radioactivity concentrations (Bq/cc) can be used to evaluate contrast in tomographically reconstructed images. The phantom image should be acquired with high information density (counts/cc) to minimize statistical noise. Initial testing of new systems is recommended to establish acceptable image contrast for the test conditions and phantom. It is common practice to use a commercial ECT phantom to test contrast as a function of the size of various spheres. Additional information concerning assessing contrast can be found in AAPM Report 52 *Quantitation of SPECT Performance*.

15.4 NOISE AND UNIFORMITY

SPECT systems are based on Anger-type gamma cameras that acquire planar projection images at each projection angle. Tomographic images are computed using filtered back-projection from the gamma camera's planar projection images. The ramp filter accentuates high frequencies where the SNR is generally lower, and it is important that both signal and noise across the camera's FOV be uniform, that is, that the SNR does not vary spatially.

The most common quality control test for gamma cameras is the uniformity test. This test involves imaging with a uniform flux of radiation across the FOV and evaluating the image uniformity. The acquired image is called a "flood" image, and uniform radiation sources are called flood sources or uniformity phantoms. Uniformity is more critical in the central FOV or CFOV, the region of diagnostic focus, so manufacturers must report expected uniformity for both the CFOV (diameter = 75% UFOV) and for the useful FOV (UFOV) with diameter 95% = FOV. The full FOV is defined as the FWHM of the flood image. For SPECT, uniformity must be consistent across the entire FOV used for reconstruction. Uniformity testing can be done with the collimator off (intrinsic test) or with the collimator on (extrinsic test). The older AAPM Report 9 *Computer Aided Gamma Camera Acceptance Testing* provides detailed methods for testing uniformity. A more recent, and excellent source of information is the NEMA publication NU 1-2012 *Performance Measurements of Gamma Cameras*.

It is important to assess spatial uniformity for each camera head. Uniformity testing acquires a uniform flux for a total of five million counts for an LFOV camera head (two million if planar imaging only). The flood image in Figure 15.3a is an example of not using the vendor supplied lead mask to avoid excessive counts at the periphery. The uniformity for this older camera was acceptable for use in planar imaging but might lead to artifacts if used for SPECT; this older system was only used for planar imaging. The multiple ROIs in Figure 15.3b are provided as examples of establishing the FOV, which was troublesome due to the excess count density at the periphery. In this case, we defined the FOV using the 50% threshold of the inner rim of this bright region (FOVin). The UFOV and CFOV were based on FOVin. To quantify spatial variance due to the camera nonuniformity for each of these ROIs, we must remove the variance due to counting statistics, assumed to be equal to the mean value:

$$SD_{camera} = \sqrt{SD_{ROI}^2 - \overline{Mean_{ROI}}} \qquad (15.4)$$

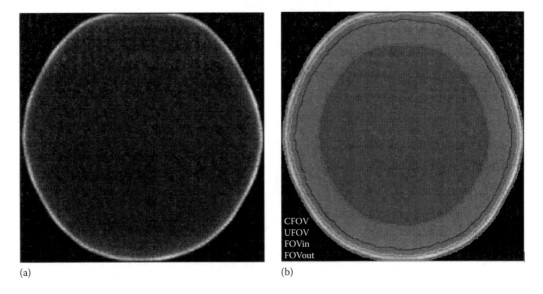

(a) (b)

FIGURE 15.3 **(See color insert.)** (a) Flood image for LFOV camera head with (b) ROIs for the FOV (outer and inter-ridge), UFOV, and CFOV.

An index of uniformity is the ratio of camera standard deviation (SD_{camera}) divided by the mean for each ROI. This measure indicated acceptable uniformity indices of 3.1% for the CFOV, where it should be <5%, and 7.9% for the UFOV, where it should be <10%. The uniformity index was excessive for FOVin (16.0%) and FOVout (46.1%). These measures of uniformity emphasize the need to focus diagnostic attention to the CFOV and UFOV of older gamma cameras.

The statistical approach based on the SD in large FOVs is not optimal to detect nonuniformity in small regions, so a method was proposed by NEMA to deal with this based on an analysis of maximum and minimum values in a moving window that is used to assess "regional" uniformity.

All modern gamma cameras use microprocessors for uniformity correction (Figure 15.4). Uniformity correction microprocessors store a mapped array of correction factors spanning the FOV, and corrections are applied on the fly as the image is acquired. Early versions of microprocessors approached correction using a postprocessing sensitivity map, scaling low-count areas up and high-count areas down. However, this method failed to maintain uniform noise levels and the SNR varied across the UFOV.

Uniformity correction microprocessors were subsequently introduced with multiple correction methods to improve results: energy correction, linearity correction, and sensitivity correction. These correction methods were based on physical principles derived from two fundamental sources of gamma camera nonuniformity: (1) spatial variations in energy resolution and (2) spatial nonlinearity. Energy resolution varies between vs. beneath PMTs and at the edge of the FOV due to differences in light collection efficiency

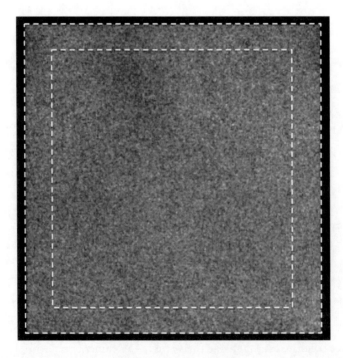

FIGURE 15.4   Modern gamma camera used for SPECT. FOV and UFOV are highlighted.

Energy resolution is based on the number of light photons collected per scintillation. Newer systems deal with the spatially varying energy resolution by varying gain and window width by location. The gain is adjusted such that the location of the peak in the pulse height spectra (called photopeak) is consistent across the FOV. Where fewer light photons are collected, the width of the distribution about the photopeak in the pulse height spectra changes, wider for fewer light photons. This is corrected by adjusting local energy window widths to match the same fraction of the distributions counts across the FOV. When properly calibrated, energy correction can provide a consistent count rate across the UFOV, such that quantum noise is uniform.

However, energy correction alone does not resolve the other fundamental source of non-uniformity in the gamma camera images, spatial nonlinearity. Spatial nonlinearity is seen as mispositioning of the site where gammas interact within the crystal. The energy from each gamma absorbed within the crystal is converted to a large number of light photons promptly emitted within the crystal (collectively called scintillation). These light photons are converted to electrical signals by the PMTs, and a weighted signal calculated based on the PMT's x- and y-location within the FOV. The individual x- and y-signals from the PMTs are summed and normalized to estimate the x- and y-location of the scintillation. Positional errors vary beneath versus between PMTs, such that parallel lines tend to bow toward PMTs. In newer systems, this source of spatial nonlinearity is evaluated by imaging a phantom with holes of fixed spacing and known positions and calculating x- and y-shifts corrections as a function of raw x and y positions. Microprocessor-based linearity corrections combined with energy corrections provide very good intrinsic uniformity (without collimator) (Figure 15.4).

The final correction provided by microprocessors is sensitivity correction, which is used to deal with minor differences in transmission across the UFOV of collimators and/or residual nonuniformity due to the camera crystal. If these minor sensitivity changes are not accounted for, ring-like artifacts can show up in reconstructed SPECT images and will be worse near the center of the reconstructed image. The combination of energy, linearity, and sensitivity corrections can provide an excellent uniformity for modern gamma cameras. The uniformity index for a modern gamma camera used for SPECT was 3.5% for the UFOV and only 3.7% for the full FOV (Figure 15.4). This gamma camera has an acceptable uniformity index for the entire FOV, which makes it ideal for SPECT. Since the SNR for a gamma camera is calculated as $N^{1/2}$, ensuring good uniformity also ensures a consistent SNR across the cameras FOV.

It is good practice to compute uniformity at a variety of locations within the imaging FOV to ensure that noise sources other than quantum noise are minimal. This could be done using ROIs such as those in Figure 15.1.

15.5 SENSITIVITY

The sensitivity of each camera head is the count rate divided by disintegration rate (CPM/mBq) and should be the same, within statistical uncertainty. Sensitivity should be evaluated with collimators used for SPECT imaging. Sensitivity for both point and plane sources should be determined. The net count rate based on total counts and a background image acquired over the same time period should be used to correct both point and plane source measures of CPM. An energy window of ±10% of the peak energy is normally used (±14 keV for Tc-99(m)).

Plane source sensitivity is measured with a uniform source larger than the collimator FOV placed at the surface of the collimator. The specified source activity must be adjusted by the ratio of FOV area to source activity area. Plane source uniformity testing can be done as part of uniformity testing if results are reported as CPM/mBq. Differences in plane source sensitivity are mainly due to spectrometer differences, NaI(Tl) crystal differences, and collimator differences. Collimators can be swapped between camera heads to test this, if supported by manufacturer.

Point source sensitivity varies with distance from the collimator. Therefore, it should be evaluated at several distances (0, 10, and 20 cm from collimator), and usually done at the center of the FOV. Marker sources using Co-57 or small sources of Tc-99(m) can be used. The activity of the Co-57 marker source (often <100 uCi) can be calculated from its calibration data (activity with date) and half-life. The activity of the Tc-99(m) source (often >1 mCi) can be determined using a dose calibrator. While the size of a point source's image increases with distance, the system count is nearly constant at least up to 20 cm. Since image size varies with distance, the count rate from the point source needs to be assessed with an ROI with size adjusted to match the change in point source size. This can be done using an ROI with bounds set to 1%–5% of the peak value in the image. The net count within the ROI after background subtraction should be greater than 10,000. The net count is determined by subtracting the ROI background count from the raw count, where the background count is assessed for the same time but without the point source present. Results may be expressed in terms of CPM/uCi or CPM/mBq.

HOMEWORK PROBLEMS

P15.1 Determine the uniformity indices for CFOV and UFOV for the planar flood image provided.

P15.2 Estimate the spatial resolution as SD-based contrast for each the bar pattern in the bar phantom image provided.

P15.3 The QC tech reports that the plane source sensitivity on one of the gamma cameras in the department has dropped by more than 10% in 1 day. What camera problems could cause this? What operator errors might lead to this apparent loss in sensitivity? How would you test for these?

CHAPTER 16

Magnetic Resonance Imaging

16.1 INTRODUCTION

Unlike other medical imaging systems presented in this book, magnetic resonance imaging (MRI) does not involve ionizing radiation. Though MRI produces tomographic images, the use of terms such as emission-computed tomography (ECT) and transmission-computed tomography (TCT) is not appropriate, and these categories are best left to imaging systems that employ ionizing radiation. Therefore, it seems reasonable to simply categorize MRI as CT, since we are dealing with RF radiations that are both transmitted into the body as well as those that are subsequently emitted from the body (rather than something more confusing like transmission and emission computer tomography (TECT).

MR images are acquired slice by slice or by encoding the entire volume. I will use the slice-by-slice approach common to the spin-echo-style imaging to simplify the description. A slice is selected using a slice-encoding gradient that slightly alters the magnetic field and frequencies in the slice direction. The slice-encoding gradient takes a short time to stabilize when turned on. Then, an RF signal is transmitted into the body to excite spins at the desired slice position (based on center frequency) and thickness (based on bandwidth). After desired excitation, the RF and slice-encoding gradient are turned off. The timing between RF transmissions and timing for reception of RF signals from the slice provide the basis for contrast between tissues (Section 4.7.2). During reception, RF signals from the selected slice are spatially encoded with frequencies that vary across the slice using a second gradient (position-encoding gradient). A mixer circuit in the RF receiver removes the high-frequency carrier signal (~43 mHz at 1 T), resulting in a position-encoded lower-frequency signal (e.g., −16 to +16 kHz). An ADC samples this lower-frequency position-encoded signal, and the samples are stored as values in one row of a 2-D image called a k-space image (usually 256 × 256). This is repeated, applying phase encoding in the direction perpendicular to the frequency encoding direction, to change the row location to fill in the k-space array. The 2-D k-space images (actually real and imaginary parts) are the Fourier transform of the signals from the slice. An inverse Fourier transform is used to convert the 2-D k-space image to real and imaginary 2-D images of the object. The real and imaginary images are transformed to magnitude and phase images, and the

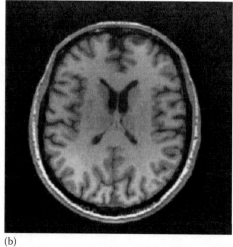

(a)   (b)

FIGURE 16.1   (a) The magnitude of k-space used to form the MR image of the head (b) as the Fourier transform of k-space. *Note:* The k-space magnitude is log10(*k*-values) to capture the range of frequencies (zero for *kx* and *ky* is at the center of the k-space image).

2-D magnitude image serves as the MR image of signals from the slice. This is repeated to acquire multiple slices forming a multislice tomographic image of the object (Figure 16.1). *Note:* Phase encoding is a much slower process than frequency encoding such that motion tends to produce artifacts in the phase-encoded direction in the final MR image.

As with all medical imaging systems, it is important to understand the origin and features of spatial resolution, contrast, and noise. I will begin with noise as its origin differs substantially from that in x-ray and nuclear imaging.

16.2 MODELING RANDOM NOISE

The aforementioned background describing MR image acquisition as a magnitude signal helps to understand the nature of the noise distribution function in MRI. Each magnitude image's pixel value is calculated as follows:

$$S_m = [(S_r + x)^2 + (S_i + y)^2]^{1/2} \tag{16.1}$$

where
  S_m is the signal magnitude
  S_r and S_i are real and imaginary signals after Fourier transform of the k-space image

Here, x and y are assumed to be zero mean additive random noise (Gaussian random variables) with identical standard deviations (σ). Expanding Equation 16.1 leads to

$$S_m = \left(S_r^2 + S_i^2 + 2S_r x + 2S_i y + x^2 + y^2 \right)^{1/2} \tag{16.2}$$

The first two terms in this equation relate to the theoretical noise-free signals arising from the real and imaginary parts. These are the desired signals, and if they are large compared to noise levels, then $S_m \sim (S_r^2 + S_i^2)^{1/2}$ with little contamination from the noise terms. A high SNR is not always possible, especially for functional MRI (fMRI) studies.

We saw when we calculated the pdf for a squared random variable in Section 7.6 that it was not Gaussian. In regions where the signal is nonzero, the pdf is quite complex. However, for cases where the signal terms can be assumed to be zero (e.g., outside the body), then the noise magnitude is $\rho \sim (x^2 + y^2)^{1/2}$ and we find that the pdf(ρ) in these areas follows a *Raleigh distribution* (Figure 16.2) with the following form:

$$\text{pdf}_m(\rho) = \frac{\rho}{\sigma^2} e^{-\frac{1}{2}\left(\frac{\rho}{\sigma}\right)^2} \tag{16.3}$$

where

ρ is noise magnitude expressed in 2-D polar coordinates (ρ, $\theta = 0$)
σ is its scale parameter

Using pdf$_m$(ρ) from Equation 16.3, we can determine the following relationships:

$$\text{Mean value of noise magnitude} = \mu_\rho = \sigma\sqrt{\frac{\pi}{2}} \approx 1.25\sigma. \tag{16.4}$$

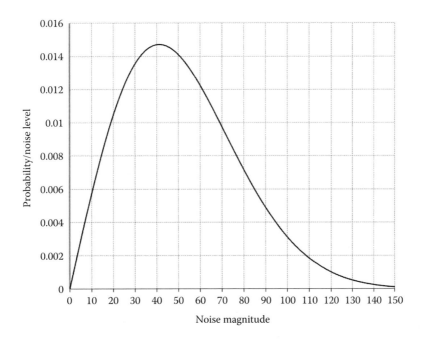

FIGURE 16.2 Raleigh distribution with scale parameter ($\sigma = 28$) determined for an ROI in the background of anatomical MR image of brain from Figure 16.1b.

$$\text{Standard deviation of noise magnitude} = \left(2 - \frac{\pi}{2}\right)^{1/2} \sigma \approx 0.655\sigma. \tag{16.5}$$

The scale parameter σ can be evaluated using an ROI outside the object, usually in one corner of the image. *Caution: do not include noise due to phase artifacts.* If we measure the mean value of the region according to Equation 16.4, it should be $\sim 1.25\sigma$. Alternatively, if we measure the standard deviation within the region according to Equation 16.5, it should be 0.665σ. If all is well both approaches should lead to the same value for this measure of the standard deviation of the noise.

Signal-to-noise ratio (SNR): There is a problem when attempting to calculate the SNR, since signal magnitude also includes noise terms (Equation 16.1). One approach to deal with this problem and to estimate SNR is based on the analysis of an image of a uniform cylindrical phantom. A large region of interest (ROI) can be placed over the phantom image, and the mean value of this ROI can be used to estimate the signal. In this case, the ROI mean value is a good approximation of the true signal since the ROI should contain several hundred pixels, and the average of the mixed signal–noise terms in Equation 16.2 should therefore average toward zero. Variance in a signal-void ROI can be used to calculate the noise standard deviation. SNR can be estimated using the signal determined from the mean of the signal magnitude within the phantom corrected for noise magnitude estimated from the signal-void region.

16.2.1 Tissue-to-Tissue SNR

The aforementioned method is good for specifying SNR of an object relative to a zero signal, as is done when specifying system SNR by manufacturers, but the SNR of importance for medical imaging relates to low-contrast signals and is the focus of most signal detection problems in imaging. As described in Chapter 1, this low-contrast SNR is based on the signal difference between adjacent tissues and associated noise fluctuations. For practical purposes, we can draw ROIs in adjacent tissues to estimate tissue-to-tissue SNR. This SNR is calculated using the mean (μ) and standard deviation (σ) from each tissue ROI as follows:

$$\text{SNR}_{\text{tissue 1-tissue 2}} = \frac{\mu_{\text{tissue 1}} - \mu_{\text{tissue 2}}}{\left(\sigma_{\text{tissue 1}} + \sigma_{\text{tissue 2}}\right)/2} \tag{16.6}$$

The ROIs should be approximately the same size for both tissues. Equation 16.6 is consistent with the definition of tissue SNR in Chapter 1 and takes into account the inhomogeneity of the tissues in its noise estimate. This measure is based on an estimate of the average standard deviations for the two tissues.

16.2.2 Noise Bias in Estimating T1 and T2

An important issue arises when determining T1 or T2 from relaxation data at numerous time points along a signal magnitude relaxation curve $S_m(t)$. This is particularly

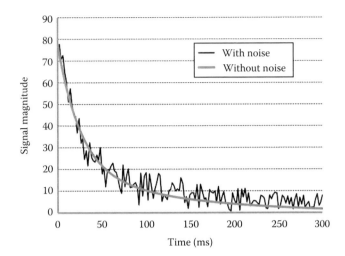

FIGURE 16.3 Two-compartment T2 relaxation curve with and without random noise (in compartment 1, T2 = 25 ms; in compartment 2, T2 = 120 ms). Signal from compartment "1" was ~2–2/3 that from compartment "2."

important for T2 where we want to sample early and late time points in the relaxation curve and may be modeling relaxation as a multiexponential process (Figure 16.3). From Equation 16.2, we see that the random noise that is added to the magnitude signal due to the x^2 and y^2 terms is always positive, and this leads to a positive bias relative to the true signals, that is, the measured signal magnitude will be larger than the true signal magnitude. Interestingly, if we fit the square of Equation 16.2 (sometimes called the power), we can correct for this bias by assuming noise variance (ρ^2) is constant, which is to be determined during the fit. This removes the bias since the expected value of $\rho^2 = (x^2 + y^2)$ should be the same at each time point. Potential bias due to the cross terms in Equation 16.2 will be minimized by the fitting process as x and y are approximately zero mean. The resulting fit can be used to determine the corrected relaxation curve, and T2 values are determined.

16.3 SPATIAL RESOLUTION

The concern with spatial resolution in clinical MRI is for those pulse sequences that need to provide resolutions of 1 mm or less. These are usually acquired with a 3-D-style acquisition and a gradient echo technique.

16.3.1 Readout Direction

There are several factors that affect spatial resolution in the readout direction. During readout, a magnetic field gradient is applied to encode position using a range of frequencies about the Larmor frequency. The Larmor frequency is removed in the RF receiver, leaving the encoded frequency range of $-f_{max}$ to $+f_{max}$. This range is selected to cover the desired FOV for the readout direction. The maximum frequency is that supported by the sampling rate of the ADC (f_s), and according to Shannon's sampling theorem, $f_s \geq 2f_{max}$ to avoid aliasing. Following are data for two commercial systems:

System 1

ADC (sampling rate) = 33.333 kHz   Dwell time = 30 us
G_{max} = 45 mT/m   Max FOV = 50 cm

System 2

ADC (sampling rate) = 50 kHz   Dwell time = 20 us
G_{max} = 45 mT/m   Max FOV = 55 cm

Such specifications change periodically, so they should only be considered examples. The maximum FOV is based on the magnet bore size with gradients and RF coils in place. The dwell time is the time between samples by the ADC, so it is just the inverse of the ADC's sampling rate.

The image field of view (FOV) in the readout direction is based on the range of frequencies supported by the ADC (±16.67 kHz for System 1, or bandwidth BW = 33.33 kHz), and the gradient strength is adjusted to produce this range about the Larmor frequency. During setup for acquisition, the gradient strength is set to select the desired FOV with smaller gradient strengths providing larger FOVs and larger gradient strengths for smaller FOVs. The readout FOV_x is determined from the following equation:

$$FOV_x\,(m) = \frac{BW\,(Hz)}{\gamma\,(Hz/T)G_x\,(T/m)} \tag{16.7}$$

This equation shows that FOV_x (in the readout direction) is inversely related to the readout gradient's strength, since BW (fixed by ADC) and γ are constant. If the FOV is smaller than the object, there will be frequencies present above the ADC's Nyquist limit (16.67 kHz in this example) and these will be aliased to lower frequencies. Aliasing in the readout direction causes the object outside of the image's FOV to appear on the opposite end of the reconstructed image (Figure 16.4). As noted in other chapters, a low-pass analog filter is used to reduce the magnitude of frequencies beyond the BW of the ADC prior to digitization. The success of such antialiasing filters depends on the strength of signals arising outside of the FOV.

Within plane pixel spacing (ΔX or ΔY) is calculated as follows:

$$\Delta X = FOV_x/N_x$$

$$\Delta Y = FOV_y/N_y$$

where FOVs are in mm and N_x and N_y are the number of samples in the frequency encode (x) and phase encode (y) directions. For a 25 cm FOV (used for brain imaging) with $N_x = N_y = 256$, the pixel spacing is slightly less than 1 mm. Since dwell times are very small, the readout of 256 samples is completed in less than 10 ms. However, the time between

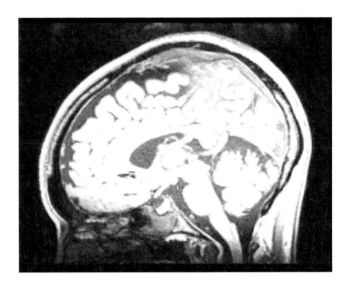

FIGURE 16.4 Note the aliased nose and neck appearing at the top of the head. This aliasing is attenuated strongly and does not interfere with the brain. The display window and level settings were adjusted to show the weaker signals from the aliased signals.

samples in the phase encode direction is >10 ms/sample, and with 256 samples, the total time to complete phase encoding is >2.5 s in this example.

For System 1 with $N_x = 256$, the bandwidth/pixel abbreviated as BWpp = 33.333 kHz/256 pixels ~130 Hz/pixel. The BWpp must be large compared with magnetic field inhomogeneity to ensure that tissue signals fall within the intended voxel.

16.4 CONTRAST

We covered T1-weighted, T2-weighted, and proton density–weighted signals and their contrast in Chapter 4. However, MRI pulse sequences can be manipulated in a variety of ways to produce images with differing contrast. The following are discussions of contrast mechanisms for several other important pulse sequences used in MR imaging.

16.4.1 Blood Oxygen Level–Dependent Weighting

Blood oxygen level–dependent (BOLD) weighting is used in functional brain MRI or fMRI studies. When neurons are active, they change the arterial hemoglobin from oxyhemoglobin to deoxyhemoglobin, increasing its magnetic moment and the local magnetic susceptibility. This reduces the local tissue spin–spin relaxation times (T2 and T2*) potentially reducing BOLD MRI signals. However, there is also vasodilation and increased blood flow that offsets the increase in deoxyhemoglobin by delivering additional oxyhemoglobin. The end result is that BOLD pulse sequences, which are susceptibility weighted, have increased signals in the brain where there is more neural activity. Acquiring a series of images with and without planned stimulus of brain activity provides a means to formulate images where activated brain areas are significantly different from nonactivated areas. These images are formulated as statistical parametric images, and when fitted to a standard brain atlas, they are called statistical parametric maps or SPMs.

16.4.2 Diffusion Tensor Imaging

Diffusion tensor imaging (DTI) provides a means to assess the magnitude and direction of water molecule diffusion *in vivo*. Since water diffusion is multidirectional, we attempt to model this using a 3-D ellipsoid model. The ellipsoid model has three major axes (represented by three orthogonal vectors) and three associated diffusion magnitudes for each voxel within a 3-D image, usually spanning the brain. The tensor nature of the ellipsoidal model leads to an image representation problem. One approach calculates scalars from the ellipsoid model to use as voxel values. The most common scalars are axial, radial, and mean diffusivity and fractional diffusion anisotropy. Axial diffusivity is along the major axis of the ellipse, radial diffusivity is the average of the two smaller axial directions, and mean diffusivity is the average of all three. Fractional anisotropy (FA) is a measure of asymmetry of water diffusion with values ranging from near zero for symmetric diffusion to near unity where diffusion is much longer along the major axis. FA tends to be larger in more heavily myelinated white matter tracts and is often assumed to follow myelin levels, but other methods are needed to ensure this.

In addition to producing a variety of scalar signal images for DTI, we are challenged to represent directions of the major axis of diffusion in images. This has led to displays using 3-D graphics where voxels have been linked based on similar directions to mimic nerve tracts, a method called tractography. The linked voxels are formed into tube-like objects for graphical display and analysis. DTI imaging is mostly used to assess white matter integrity in the brain. More complex models are now being used to deal with limitations of the assumptions of the ellipsoidal model in the complex white matter structure within the brain.

16.4.3 Arterial Spin Labeling

Arterial spin labeling (ASL) in MRI is similar to blood flow studies in nuclear medicine. In ASL studies, arterial blood is labeled by altering its magnetization, and the difference between labeled and unlabeled signals is used to estimate tissues' blood flow downstream. The advantage is that no ionizing radiation is needed and the assessment can be repeated many times under a variety of conditions to assess changes in blood flow. ASL is gaining popularity for use in brain functional imaging studies.

Contrast in these special MR procedures is not as important as the uniqueness of their signals. This does not mean that we are not interested in having high contrast between tissues of interest. In MR imaging, the pulse sequences can be chosen to provide signals that are more physiologically based than options for acquisition parameters for x-ray images. Additionally, MR imaging sessions often provide several types of MR images with pulse sequence weighting selected to optimize diagnostic and/or research objectives.

HOMEWORK PROBLEM

P16.1 Use the Equation 16.3 given for the Raleigh distribution to derive the equations for mean and standard deviation of the noise (Equations 16.4 and 16.5). Show that the peak value of the Raleigh distribution given in Equation 16.3 occurs at $\rho = \sqrt{2}\sigma$.

Solution:

Raleigh equation

(a) Mean value

$$\langle\rho\rangle = \int_0^\infty \rho \cdot \text{pdf}(\rho)d\rho = \int_0^\infty \rho \cdot \frac{\rho}{\sigma^2} e^{-\frac{1}{2}(\rho/\sigma)^2} d\rho$$

$$\text{let } k = \frac{\rho}{\sigma} \text{ so } \rho = k\sigma \text{ and } d\rho = \sigma dk$$

$$\langle\rho\rangle = \sigma \int_0^\infty k^2 e^{-\frac{1}{2}k^2} dk$$

Integral from mathtables leads to a mean value of

$$\langle\rho\rangle = \sigma \cdot \left(\frac{\sqrt{2\pi}}{2}\right) = \sigma \cdot \sqrt{\frac{\pi}{2}}$$

Standard Deviation

$$Var(\rho) = \langle\rho - \bar{\rho}^2\rangle = \langle\rho^2 - 2\rho\bar{\rho} + \bar{\rho}^2\rangle$$

$$= \langle\rho^2\rangle - 2\bar{\rho}^2 + \bar{\rho}^2 = \langle\rho^2\rangle - \bar{\rho}^2 = \langle\rho^2\rangle - \frac{\pi}{2}\sigma^2$$

now

$$\langle\rho^2\rangle = \int_0^\infty \rho^2 \text{pdf}(\rho)d\rho = \int_0^\infty \rho^2 \frac{\rho}{\sigma^2} e^{-\frac{1}{2}(\rho/\sigma)^2} d\rho$$

$$\text{let } k = \frac{\rho}{\sigma} \text{ and } d\rho = \sigma dk$$

$$\langle\rho^2\rangle = \sigma^2 \int_0^\infty k^3 e^{-\frac{1}{2}k^2} dk$$

from mathtables for the integration

$$\langle\rho^2\rangle = 2\sigma^2$$

so

$$Var(\rho) = 2\sigma^2 - \tfrac{\pi}{2}\sigma^2 = \sigma^2\left(2 - \tfrac{\pi}{2}\right)$$

so the standard deviation is

$$SD = \left(2 - \tfrac{\pi}{2}\right)^{\frac{1}{2}}\sigma$$

(b) Peak value of probability density function occurs where first derivative = 0.

$$\text{pdf}(\rho) = \tfrac{\rho}{\sigma^2}e^{-\frac{1}{2}(\rho/\sigma)^2}$$

$$\frac{d}{d\rho}(\text{pdf}(\rho)) = \frac{d}{d\rho}\left(\frac{\rho}{\sigma^2}\right)e^{-\frac{1}{2}(\rho/\sigma)^2} + \frac{\rho}{\sigma^2}\frac{d}{d\rho}\left(e^{-\frac{1}{2}(\rho/\sigma)^2}\right) = 0$$

$$\frac{1}{\sigma^2}e^{-\frac{1}{2}(\rho/\sigma)^2} + \frac{\rho}{\sigma^2}\left(\frac{-2\rho}{2\sigma^2}\right)e^{-\frac{1}{2}(\rho/\sigma)^2} = 0$$

$$1 - \frac{\rho^2}{\sigma^2} = 0$$

so $\rho = \sigma$

=0 since both must be positive.
This can be verified by graphing the pdf in excel and is consistent with graph in the chapter.

P16.2 A problem involving relaxometry to estimate T1 and T2 relaxation times (see excel file).

Index